Beginner's Guide to
Reading
Schematics

Third Edition

Stan Gibilisco

Mc
Graw
Hill
Education

New York   Chicago   San Francisco   Athens   London
Madrid   Mexico City   Milan   New Delhi
Singapore   Sydney   Toronto

Beginner's Guide to Reading Schematics, Third Edition

2 3 4 5 6 7 8 9 0   DOC/DOC   1 9 8 7 6 5

ISBN 978-0-07-182778-2
MHID 0-07-182778-1

This book is printed on acid-free paper.

Sponsoring Editor Judy Bass	**Project Manager** Nancy Dimitry, D&P Editorial Services	**Proofreader** Don Dimitry, D&P Editorial Services
Editing Supervisor Stephen M. Smith	**Copy Editor** Nancy Dimitry, D&P Editorial Services	**Art Director, Cover** Jeff Weeks
Production Supervisor Pamela A. Pelton		**Composition** D&P Editorial Services
Acquisitions Coordinator Amy Stonebraker		

About the author

Stan Gibilisco has authored or coauthored more than 50 books on physics, electronics, mathematics, and computing. He began his career as an electronics engineer, wireless broadcast technician, and magazine editor. The American Library Association named Stan's *McGraw-Hill Encyclopedia of Personal Computing* (1995) as one of the "Best References of 1996." In addition to authoring several books in McGraw-Hill's *DeMYSTiFieD* series of home study guides, he has written three books in McGraw-Hill's *Know-It-All* series for students who enjoy mathematics. Stan's *Teach Yourself Electricity and Electronics* has become a classic in the field.

Contents

Introduction   ix

1   What's the scheme?   1
Block diagrams   2
Schematic diagrams   2
Schematic symbology   4
Schematic interconnections   6
Visual language   8

2   Block diagrams   13
A simple example   13
Functional drawings   13
Current and signal paths   16
Flowcharts   18
Process paths   22
Summary   24

3   Component symbols   25
Resistors   25
Capacitors   30
Inductors and transformers   34
Switches   39
Conductors and cables   43
Diodes and transistors   46

Electron tubes   49
Cells and batteries   53
Logic gates   54
Summary   56

4   Simple circuits   *57*
Getting started   57
Component labeling   66
Troubleshooting with
  schematics   71
A more complex circuit   75
Schematic/block combinations   78
Summary   81

5   Complex circuits   *83*
Identifying the building blocks   83
Page breaks   91
Some more circuits   94
Getting comfortable with large
  schematics   106
Summary   111

6   Let's learn by doing   *113*
Your breadboard   113
Wire wrapping   117
Kirchhoff's current law   119
Kirchhoff's voltage law   123
A resistive voltage divider   125
A diode-based voltage reducer   132
Mismatched lamps in series   137
Summary and conclusion   144

A   Schematic symbols   *145*

B   Resistor color codes   *163*

**Suggested additional
  reading   167**

Index   169

Introduction

Have you "caught the electronics bug," only to grow sick with apprehension as you encountered diagrams with arcane symbols the moment you decided to build, troubleshoot, or repair something? If so, you're holding the cure in your hands right now.

A little knowledge of electronics symbology can eliminate a lot of fear and confusion. Don't give up electronics just because you've come across some spooky graphics. That would be like giving up a sport because you fear the pain of training. That's where the coach comes in! A good coach streamlines your training in any sport and helps you get past the pain. Schematic diagrams, well drawn and clearly portrayed, can help you to design, build, maintain, troubleshoot, and repair electronic equipment.

Schematic diagrams are like road maps of electronic highway networks. These drawings can help you find your way through the elements of simple circuits, complex devices, and massive systems. Once you've learned what the symbols and notations stand for, reading a schematic will come as easily to you as planning a trip with the aid of a road atlas.

This book contains all the information that you'll need in order to begin exploring electronic circuits. It can help you build a solid background for a career in electronics, whether you decide to go into design, maintenance, or repair. This book explains the rationale behind schematic diagrams, how to use or interpret each symbol, and how the symbols interconnect to reflect real-world circuits. You'll even get a chance to diagram and perform some experiments!

Once you've completed this book, you'll have plenty of information and confidence, so that you can continue your quest to enter whatever field of electronics suits your fancy, whether it's something as humble as shortwave or amateur radio, or something as cutting-edge and exotic as bioelectronics, space communications, or mechatronics.

I welcome your suggestions for future editions. Please visit my website at www.sciencewriter.net. You can e-mail me from there.

Above all, have fun!

Stan Gibilisco

1

What's the scheme?

You'll encounter three types of diagrams in electricity and electronics: block, schematic, and pictorial. Each type of diagram serves its own special purpose.

1. A block diagram gives you an overview of how the discrete circuits within a device or system interact. Each circuit is represented with a "block" (a rectangle or other shape, depending on the application). Interconnecting lines, sometimes with arrows on one or both ends, reveal the relationships between the circuits.

2. A schematic diagram (often simply called a schematic) includes every component that a circuit contains, with each component having its own special symbol. This book is devoted mostly to schematics.

3. A pictorial diagram, sometimes called a layout diagram, shows the actual physical arrangement of the circuit elements on the circuit board or chassis, so that you can quickly find and identify components to test or replace.

When you troubleshoot an unfamiliar electronic circuit, you'll usually start with the block diagram to find where the trouble originates. Then you'll refer to the schematic diagram (or part of it) to find the faulty component in relation to other components in the circuit. A pictorial diagram can then tell you where the faulty component physically resides, so that you can test it and, if necessary, replace it.

Block diagrams

Block diagrams work well in conjunction with schematics to aid circuit comprehension and to streamline troubleshooting procedures. Each block represents all of the schematic symbols related to that part of the circuit. In addition, each block has a label that describes or names the circuit it represents. However, the block does nothing to explain the actual makeup of the circuit it represents. The blocks play a functional role only; they describe the circuit's purpose without depicting its actual components. Once you've gained a basic understanding of the circuit functions by looking at the block diagram, you can consult the schematic for more details.

To understand how you might use block diagrams, consider the following two examples.

First, suppose that you want to design an electronic device to perform a specific task. You can simplify matters by beginning with a block diagram that shows all of the circuits needed to complete the project. From that point, you can transform each block into a schematic diagram. Eventually, you'll end up with a complete schematic that replaces all of the blocks.

Alternatively, you can go at the task the other way around. Imagine that you have a complicated schematic, and you want to use it to troubleshoot a device. Because the schematic shows every single component, you might find it difficult to determine which part of the device has the problem. A block diagram can provide a clear understanding of how each part operates in conjunction with the others. Once you've found the troublesome area with the help of the block diagram, you can return to the schematic for more details.

Schematic diagrams

A schematic diagram acts, in effect, as a map of an electronic circuit, showing all of the individual components and how they interconnect with one another. According to one popular dictionary, the term *schematic* means "of or relating to a scheme; diagrammatic." Therefore, you can call any drawing that depicts a scheme—electronic, electrical, physiological, or whatever—a schematic diagram.

One of the most common schematic diagrams finds a place in almost every car or truck in the United States. It's a road map, of course, and it portrays a specific sort of scheme. The scheme might involve the paths of travel within a small town, within a state or province, or across multiple states or provinces. Like a schematic diagram of an electronic circuit, the road map shows all the components relevant to the scheme it addresses. Motorists make up their own schemes, which often comprise small portions of the total scheme included in the road map. Likewise, an electronic schematic shows all of the relevant components, and it allows a technician to extrapolate the components and interconnections when testing, troubleshooting, and repairing a small circuit, a large device, or a gigantic system.

Imagine that you want to travel in your automobile from point A to point B. Your road map lists all of the towns and cities that lie between these two points. By comparison, a schematic diagram lists all the components between a similar point A and point B in an electronic circuit. Nevertheless, both of these schematics indicate much more than mere points. You need to know more than which towns or cities lie between two fixed points to get an idea of the overall scheme of things. Indeed, you could easily write down the names of various towns or landmarks, in which case you would not have to resort to a road map at all. From an electronics standpoint, you could do the same thing by compiling a list of the components in a certain circuit, such as:

- 120-ohm resistor
- 1000-ohm resistor
- PNP transistor
- 0.47-microfarad capacitor
- 2 feet of hookup wire
- 1.5-volt battery
- Switch

This list tells us nothing about the circuit in a practical, put-together sense. We know all of the components that we would need if we wanted to build the thing, but we don't know what it would be if we did! In fact, these components might go together in two or three different ways to make two or three different circuits with different characteristics.

A schematic drawing must indicate not only all components necessary to make a specific scheme, but also how these components interrelate to one another. The road map connects various towns, cities, and other trip components with lines that represent streets and highways. A line that indicates a secondary road differs from a line that represents a four-lane highway. With practice, you can learn to tell immediately which types of lines indicate which types of roads. Likewise, an electronic schematic drawing uses a plain, straight line to indicate a standard conductor; other types of lines represent cables, logical pathways, shielding components, and wireless links. In all cases, when you draw the interconnecting lines, you draw them in order to indicate relationships between the connected components.

Schematic symbology

A schematic diagram reveals the scheme of a system by means of *symbology*. On a map, the lines that indicate roadways constitute symbols. But of course, a single black line that portrays Route 522 in no way resembles the actual appearance of this highway as we drive on it! We need know only the fact that the line symbolizes Route 522. We can make up the other details in our minds. If people always had to see pictorial drawings of highways on paper road maps, those maps would have to be thousands of times larger than those folded-up things we keep in our vehicle glove compartments, and they would be impossible for anybody to read.

Tip
Since the previous edition of this book was published in 1991, portable computers and the Internet have evolved so that, today, you actually can find and access road maps that show pictorial drawings of some roads and highways! You can look at photographs taken from satellites, aircraft, and sometimes even vehicles that have driven along specific routes. Check out "Google Maps," for example. These maps aren't on paper; instead, they reside in cyberspace. You need a computer or tablet device to use such "supermaps," but they do exist, and they're getting better every day.

On a decent road map, you'll find a key to the symbols used. The key shows each symbol and explains in plain language what each one means. If a small airplane drawn on the map indicates an airport and you know this fact, then each time you see the airplane symbol, you'll know that an airport exists at that particular site, as shown on the map. Symbology depicts a physical object (such as an airport outside a large city) in the form of another physical object (such as an airplane image on a piece of paper).

A good road map contains many different symbols. Each symbol is human engineered to appear logical to the human mind. For instance, when you see a miniature airplane on a road map, you'll reasonably suppose that this area has something to do with airplanes, so a detailed explanation should not be necessary. If, on the other hand, the map maker used a beer bottle to represent an airport, anyone who failed to read the key would probably think of a saloon or liquor store, not an airport! Because a map needs many different symbols, a good map maker will always take pains to make sure that the symbols make logical sense.

Pure logic will take us only up to a certain point in devising schemes to represent complicated things, especially when we get into the realm of electronic circuits and systems. For example, a circle forms the basis for a transistor symbol, a light-emitting-diode (LED) symbol, a vacuum tube symbol, and an electrical outlet symbol. Additional symbols inside the circle tell us which type of component it actually represents. A transistor is an active device, capable of producing an output signal of higher amplitude than the input signal. We can say the same thing about a vacuum tube, but not about an LED or an electrical outlet.

A circle with electrode symbols inside has been used for many years to represent a vacuum tube. Transistors were developed as active devices to take the places of vacuum tubes, so the schematic symbol for the transistor also started with a circle. Electrode symbols were inserted into this circle as before, but a transistor's elements differ from a tube's elements, so the transistor symbol has different markings inside the circle than the tube symbol does. The logic revolves around the circle symbol. Transistors accomplish many of the same functions in electronic circuits as vacuum tubes do (or did), so symbolically they are somewhat similar.

Inconsistencies arise in schematic symbology, and that's a bugaboo that makes electronics-related diagrams more sophisticated than road maps. A circle can make up a part of an electrical symbol for a device that doesn't resemble a tube or transistor at all. An LED, for example, can be portrayed as a circle with a diode symbol inside and a couple of arrows outside. An LED is not a transistor or tube, and the electrode symbol at the center clearly reveals this difference. An electrical outlet can serve as another example. It's absolutely nothing like a tube, transistor, or LED! Yet the basis for the symbol is a circle, just like the circle for a tube or transistor or LED. You'll learn more about specific schematic symbols in Chap. 3.

Schematic interconnections

To further explore how schematic diagrams are used, let's consider a single component, a PNP transistor. This device has three electrode elements, and although many different varieties of PNP transistors exist, we draw all their symbols in exactly the same way. We might find a PNP transistor in any one of thousands of different circuits! A good schematic will tell us how the transistor fits into the circuit, what other components work in conjunction with it, and which other circuit elements depend on it for proper operation. A transistor can act as a switch, an amplifier, an oscillator, or an impedance-matching device. A single, specific transistor can serve any one of these purposes. Therefore, if a transistor functions in one circuit as an amplifier, you can't say that the component will work as an amplifier only, and nothing else. You could pull this particular transistor out of the amplifier circuit and put it into another device to serve as the "heart" of an oscillator.

Tip
By knowing the type of component alone, you can't tell what role it plays in a circuit until you have a good schematic diagram showing all the components in the circuit, and how they all interconnect. Rarely can you get all this information in easy-to-read form by examining the physical hardware. You need a road map—a schematic diagram—to show you all the connections that the engineers and technicians made when they designed and built the circuit.

Suppose that you plan to drive your car from Baltimore, Maryland to Los Angeles, California. Even if you've made the trip several times in the past, you probably don't recall all of the routes that you'll need to take and all of the towns and cities that you'll pass along the way. A road map will give you an overall picture of the entire trip. Because all of the trip data exists in a form that you can scan at a glance, the road map plays a critical role in allowing you to see the entire trip rather than each and every segment, one at a time. A schematic diagram does the same thing for a "trip" through an electronic circuit.

Continuing with the road map and the coast-to-coast trip as an example, imagine that you have memorized the entire route from Baltimore to Los Angeles. Assume also that one of the prime highways on the way is under construction, forcing you to take an alternate route. Without a road map, you'll have no idea as to what detours exist, which alternate route is the best one to take, and which detour constitutes a path that will keep you on course as much as possible and eventually return you to the original travel route with a minimum of delay and inconvenience.

An electronic circuit has many electrical highways and byways. Occasionally, some of these routes break down, making it necessary to seek out the problem and correct it. Even if you can visualize the circuit in your head as it appears in physical existence, you'll find it impossible to keep in your "mind's eye" all the different routes that exist, one or more of which could prove defective. When I speak here of visualizing the circuit, I don't mean the schematic equivalent of the circuit, but the actual components and interconnections, known as the *hard wiring*.

A schematic diagram gives you an overall picture of a circuit and shows you how the various routes and components interact with other routes and components. When you can see how the overall circuit depends on each individual circuit leg and component, you can diagnose and repair the problem. Without such a view, you'll have to "shoot in the dark" if you want to get the circuit working again, and you'll just as likely introduce new trouble as get rid of the original problem!

Fear not!

Look at the schematic of Fig. 1-1. If you've had little or no experience with these types of diagrams, you might wonder how you'll ever manage to interpret it and follow the flow of electrical currents through the circuit that it represents. Fear not! By the time you finish this book, assuming that you already know some basic electricity and electronics principles, you'll wonder how you ever could have let a diagram like this intimidate you. By the way, you'll see this diagram again in Chap. 5.

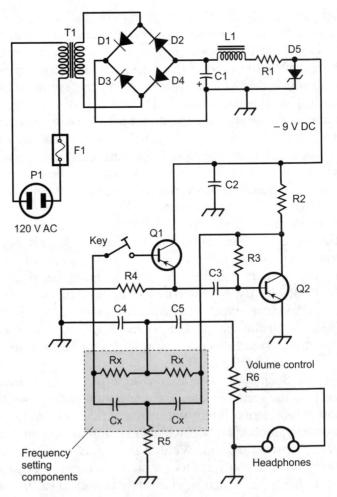

FIG. 1-1. *An example of a fairly complicated schematic diagram. By the time you finish this book, you'll find it simple!*

Visual language

It can prove difficult to explain schematic diagramming in detail to people who have just begun their study of electronics. It helps to think of this form of symbology as a language, that is to say, a system of symbols that helps us to communicate ideas. The English language (or whatever your native language happens to be) is a scheme with a system of symbology with which you're familiar, of course, because you're reading this book!

Every word spoken in English or any other verbal language is a complex symbol made from simpler elements called *characters*. Let's take the word "stop," for example. Without a reference key, this sound means nothing. A newborn infant hears noise coming out of your mouth, that's all! However, through learning the symbology from shortly after birth, this word begins to mean something because the child, who has begun learning to speak and understand, can compare "stop" to other words, and also to actions. We can even say that the word "stop" is a sort of symbology within symbology. The communicator's intent, when using the word "stop," can also be expressed by the phrase "Do not proceed further." This phrase also constitutes symbology, expressing a mental image of a desired action.

If we could all communicate by mental telepathy, then we wouldn't need language or the symbols that it comprises. Thinking happens a lot faster than speaking or writing or reading can go; and brain processes are the same from human to human, regardless of what language any particular individual employs when speaking, writing, or reading. A newborn baby speaks and understands no language whatsoever. However, whether that baby was born in the United States, South Africa, Asia, or wherever, thought processes take place.

The baby knows when it is hungry, in pain, frightened, or happy. It needs no language to comprehend these states. But the baby does have to communicate right from the start. For this reason, all newborns communicate in the same language (crying and laughing, mostly). As newborns comprehend more and more of their environment through improved sensory equipment (eyes, ears, nose, fingers), they collect more and more data. At this point the various languages come into play, with different societies using different verbal symbols to express mental processes. The human brain still carries on the same nonlinguistic thought processes as before, however, because thinking in terms of symbols would take far too much time and "brain storage."

The brain helps a human to transpose complex thoughts into a language, and vice versa, just as a computer translates programming languages into electronic impulses and vice versa. Imagine that a child is about to step in front of a speeding automobile. If the brain had to handle millions of data elements symbolically, we humans would spend all of our lives waiting for our brains to deliver the correct processed information, and that child would probably get killed before we could even begin to react. Rather, the brain scans all the data received by the sensory organs in real time and then sums it up into a single symbol for communication. A good audible (and hopefully loud) symbol in the above-mentioned case is "Stop!" You, seeing a child about to go out into heavy traffic, might shout that word and produce in the child's brain the appropriate sequence of processes.

Not all languages involve the spoken word. Have you heard of sign language, whereby the arms and hands are used to communicate ideas? If you've done any amateur (or "ham") radio communication, especially if you got your "ham" license back in the time when I got mine (the 1960s), you know about the Morse code as a set of communications symbols. In most instances, an entire communicating language of visual symbols is not as efficient for us humans as one composed of words and visual symbols combined. Using the symbol "stop" again, we can utter this word in many different ways. The word in itself means something, but the way we say it (our "tone of voice") augments the meaning. We can't do all that with the printed or displayed characters S, T, O, and P all by themselves in plain text.

We humans have arrived at some universal methods of modifying visual symbols. To many of us, the color red denotes something that demands immediate attention. Often, however, this color is used in conjunction with the visual symbol for a spoken word. Think of a "stop" sign, for example. It's red, right? Or think of a "yield" sign. It's yellow, representing something that demands attention, but in a less forceful way than the color red does.

Tip

Schematic diagrams rarely include color. Look in the back of a technical manual for an amateur radio transceiver, for example. Does the schematic have color? I'll bet you that it doesn't. (A few good magazines, however, do put color into their schematics.) A technical manual's schematic might not even have grayscale shading. Schematics resemble printed text or Morse code in this respect; we must convey a lot of information with a limited set of symbols, and we're constrained even as to the way in which we can portray and read those symbols.

Schematics don't lend themselves to any form of oral (audible) symbology either. When you see the symbol for, say, a field-effect transistor (FET) in a schematic diagram, you don't hear the paper or computer say, "Field-effect transistor, for heaven's sake, not bipolar transistor!" You have to make sure that you read the symbol correctly. If you want to build the circuit and you mistakenly put a bipolar transistor where an FET should go—maybe because you didn't look carefully enough at the schematic—you have no right to expect that the final device or system will work. Something might even burn out, so that when you recognize your error and replace the FET with a bipolar transistor, you'll have to troubleshoot the whole circuit before you can use it!

Our senses along with our central processor, the brain, render us less than proficient at mentally conceiving all of the workings of electronic circuits by dealing with them directly. Therefore, we have to accept data a small step at a time, compiling it in hardcopy form (through symbology) and providing hardcopy readout. We can liken this method to the "connect-the-dots" drawings in children's workbooks. Individually, the dots mean nothing, but once they are arranged in logical form and connected by lines, we get an overall picture. The dots' relationships to each other and the order in which they are connected tell us everything that we need to know.

The remaining chapters in this book start with the symbols for individual electronic components, then move on to simple circuits, and finally show you a few rather complicated circuits. Schematic symbols and diagrams are designed for human beings, so human logic constitutes a prime factor in determining which symbols mean what. In that respect, the creation and reading of schematic diagrams resembles mathematics, and in particular, good old-fashioned plane geometry!

Tip

Schematic diagrams are *encoded* representations of circuits, while pictorials show us the physical objects, often proportioned according to their relative size, and sometimes rendered so as to look three-dimensional by means of shading and perspective. Schematic diagrams depict circuit components as symbols only, without regard to their real-world size or shape, and in two dimensions (a flat piece of paper or computer screen), completely lacking depth or perspective.

2

Block diagrams

A block diagram portrays the general construction of an electronic device or system. A block diagram can also provide a simplified version of a circuit by separating the main parts and showing you how they are interconnected.

A simple example

Figure 2-1 is a block diagram of a device that converts alternating current (AC) to direct current (DC). The terminal at the left accepts the AC input. In sequence going from left to right, the electricity passes through the transformer, the rectifier, and the filter before arriving at the output as DC. In this case, the lines that connect the blocks do not have arrows because readers will naturally assume that the flow goes from left to right. The input terminal resides at the left-hand end of the diagram, and the output lies on the extreme right. In more complicated block diagrams, the interconnecting lines may include arrows to show which block affects which, or to indicate the general direction of signal flow when it might not otherwise be clear.

Functional drawings

Engineers and technicians employ block diagrams in various ways. Commonly, block diagrams indicate the interconnections between

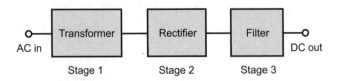

FIG. 2-1. *Block diagram of an AC-to-DC converter. The electricity flows from left to right.*

small circuits in a larger device, or between diverse devices in a large system. When drawn as shown in Fig. 2-1, block diagrams can also be called *functional diagrams* because they reveal the basic functioning of the electronic circuit. The functional diagram offers a simplistic explanation of how the device operates; it can lead to more detailed information provided by a schematic diagram.

Someone who wants to draw a schematic diagram for a complex electronic circuit designed from scratch can start with a block diagram. This diagram will show all of the circuit sections (stages) needed to arrive at a functioning device, but none of the internal details of those stages. Then the designer will develop schematic diagrams of circuits that can fill each block and serve the appropriate function in the overall system. The first block in the diagram will then be replaced by the schematic diagram of the circuit it represents. The engineer or technician will move through the blocks according to functional order, creating schematic diagrams that can be used to build each stage in the system. As soon as the final block has been filled in with a schematic for the applicable stage, the comprehensive schematic is complete, and a total (but so far only theoretical) system design is portrayed in detail.

Another way of using block diagrams starts with a finished schematic diagram. Imagine that the schematic is complicated, and that the equipment whose circuit it represents does not work properly. Although schematic diagrams can describe the functioning of an electronic circuit, they are not as clear and basic as a functional block diagram for that purpose. In the absence of a preexisting block diagram, a technician would have to start with the schematic, laboriously identify each stage in the system, and then draw the entire system diagram in block form. When finished, the block diagram would reveal how each stage interacts with the others. Using this method, one or more stages could be identified as a possible trouble area. Then the technician would refer to the original schematic and conduct tests in specific areas, based on his or her knowledge of how each stage works at the component level.

We can describe the operation of a specific type of wireless transmitter, say an *amplitude-modulated* (AM) voice transmitter, such as the type found in Citizens Band (CB) radios, by means of a block diagram. This diagram will apply to most other AM voice radio transmitters. Of course, no two transmitters built by different manufacturers are exactly alike, but all of them contain the same basic circuit sections as far as functionality goes. One type of oscillator might work differently from another type, but they all do the same thing: generate a radio-frequency (RF) signal! When we need to know, or portray, individual differences between circuits that do essentially the same things, then we need schematic diagrams.

The block diagram in Fig. 2-2 illustrates the various parts of a strobe light circuit. Let's go through the diagram block by block to understand how it works. The input signal enters at the left; it's utility AC,

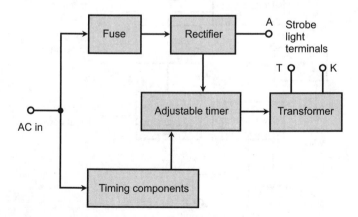

FIG. 2-2. *Block diagram of a circuit designed to provide power to a strobe light. Arrows show the direction of elecricity flow.*

such as we get from a standard wall outlet. In the United States and some other countries, this AC has a nominal voltage of 117 volts (V) and a frequency of 60 hertz (Hz), where "hertz" means "cycles per second." (In some countries, the voltage is about 234 V, and in some countries, you'll find a frequency of 50 Hz rather than 60 Hz.) The input AC goes to a fuse, and also to a combination of components that provide timing. The top path, where the fuse is located, leads to a diode-type rectifier, and the rectifier output passes directly to one terminal of the three-terminal strobe lamp. The rectifier also outputs to an adjuster that provides a variable flash rate for the lamp. The output from that adjuster goes to a transformer, which provides the remaining two outputs required to operate the lamp.

Current and signal paths

Figure 2-3 shows a power supply that produces several different voltage outputs. As you go through this diagram from left (the input) to the bottom and the right (the outputs), note that the circuit is powered with 120 volts AC (120 VAC), quite close to the nominal 117 VAC commonly found at utility outlets in the United States. The input

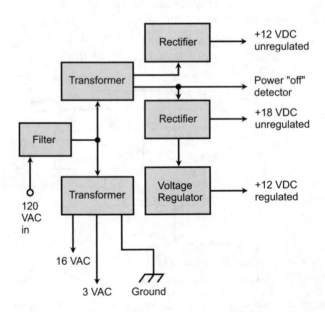

FIG. 2-3. *Block diagram of a power supply with several different outputs.*

AC goes through a filter and then splits into two paths. Part of the AC goes to the "lower" transformer that provides 16 VAC and 3 VAC output, along with a ground connection.

From the filter, the input voltage gets fed to another transformer that derives the voltages to be converted to DC electricity. One output of the transformer goes to a rectifier that provides 12 volts DC (12 VDC) without any voltage regulation. The other transformer output goes to a separate rectifier that provides 18 VDC, also unregulated. This transformer output also serves as a diagnostic detector for a power "off" condition. That line is further tapped to join with the output of the voltage regulator to provide 12 VDC with voltage regulation.

> **Tip**
> Block diagrams are comparatively easy to draw, comprising squares or rectangles along with interconnecting lines (sometimes with arrows). More sophisticated block diagrams also include triangles to represent circuit blocks built around specialized amplifiers constructed within *integrated circuits* (ICs), also known as *chips*.

Figure 2-4 is a block diagram of an AM radio transmitter. The microphone preamplifier stage goes to the input of the audio amplifier stage

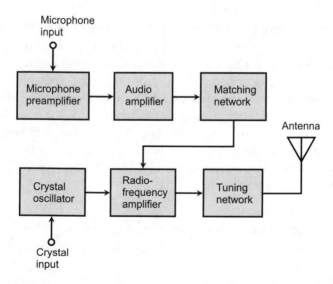

FIG. 2-4. *Block diagram of an amplitude-modulated (AM) radio transmitter.*

(note the direction of the arrow). The output of the audio amplifier goes to the matching network, which in turn goes to the RF amplifier section. The crystal oscillator is also connected to the RF amplifier section, whose output leads into the RF tuning network. Only one connection exists between the audio section of the circuit and the RF section: the one between the matching network and the RF amplifier. This block diagram, with its arrows, tells us not only how the components of the system connect to one another, but also the sequence of events or direction of signal flow.

Flowcharts

Block diagrams can describe the functioning of electronic circuits, but in the world of computers, another form of diagramming is sometimes used to portray the functioning of a program. This system is called *flowcharting*. A flowchart resembles a block diagram, except that the symbology applies to the sections of a computer program, an intangible thing (as opposed to an electronic circuit, a tangible thing). A flowchart provides a graphic representation of the logical paths that a computer will take as it executes a particular program. Flowcharts are often prepared in conjunction with specifications, and are modified as the requirements change to fit within the constraints of the computer system.

For complex problems, a formal written specification might be necessary to ensure that everyone involved understands and agrees on what the problem is, and on what the results of the program should be. To illustrate this concept, let's suppose that a teacher wants to write a computer program that will determine a student's final grade for a course by calculating an average from grades the student has received over a certain period of time. The teacher will supply the grades to the program as input. Only the average grade is needed as an output. Now, we can make an orderly list of what the program has to do:

- Input the individual grades.
- Add the grade values together to find their sum.
- Divide the sum by the number of grades to find the average grade.
- Print out the average grade.

We can prepare a flowchart of the program, as shown in Fig. 2-5. As we can see, the flowchart graphically presents the structure of the program, revealing the relationship between the steps and paths. When the flow of control is complicated by many different paths that result from many decisions, a good flowchart can help the programmer sort things out. The flowchart can serve as a thinking-out tool to understand the problem and to aid in program design. The flowchart symbols have English narrative descriptions rather than programming language statements because we want to describe *what* happens, not

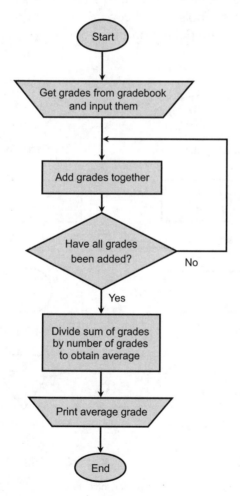

FIG. 2-5. *Example of a program flowchart.*

how it happens. At a later stage, if formal flowcharts are required for documentation, the flowchart can contain statements in a programming language. These flowcharts might prove helpful to another person who at some future time wants to understand the program.

It takes a lot of time to conceive and draw up a formal flowchart, and modifying a flowchart to incorporate changes, once a program has been written and its flowchart composed, can prove difficult. Because of these limitations, some programmers will shy away from the use of a flowchart, but for others, it can provide valuable assistance in understanding a program. In order to promote uniformity in flowcharts, standard symbols have been adopted, the most common of which are shown and defined in Fig. 2-6.

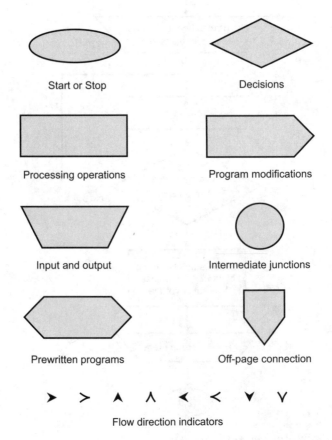

Start or Stop   Decisions

Processing operations   Program modifications

Input and output   Intermediate junctions

Prewritten programs   Off-page connection

Flow direction indicators

FIG. 2-6. *Common symbols for flowcharts intended to represent computer programs.*

Follow the flow

The normal direction of processes in a flowchart runs from top to bottom and from left to right, the same way as people read books in most of the world. Arrowheads on flow lines indicate direction. The arrows can be omitted if, but only if, the direction of flow is obvious without them.

Figure 2-7 is a flowchart for a program that duplicates punched cards, and at the same time prints the data on each card. Keep in mind that this particular "beast" is of historical interest only! (Were you born long enough ago to remember punch cards for inputting programs to computers? I recall using them, all the way back in the 1970s, when I attended the University of Minnesota. I guess that little factoid dates me, doesn't it?)

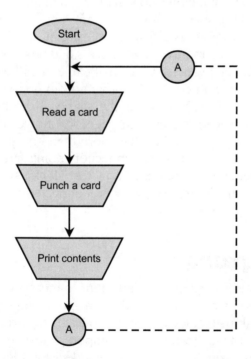

FIG. 2-7. *Flowchart for a program for duplicating punched cards. The circles labeled A represent the inflow and outflow points in the feedback loop shown by the dashed line.*

Let's trace the flow of the program through Fig. 2-7. The program begins at the "Start" oval at the top and proceeds in the direction of the arrows. In the first box below "Start," the program reads a card. Then the program punches the card's contents (data) as holes in a blank piece of heavy paper and sends the data to a printer. The program then goes back along the dashed line to the top and reads the next card. The circles marked A represent inflow and outflow points. In this case, they're superfluous, but in complicated flowcharts, they can be useful when it would create a mess to include all the applicable dashed lines. The program repeats itself as long as it has cards to read and punch.

In a sophisticated flowchart, we might see several different symbols of the sort shown in Fig. 2-6, and maybe even all of them. Oval boxes show start or stop points. Arithmetic operations go in rectangular boxes. Input and output instructions go in upside-down trapezoids. If we want to show a program that someone wrote earlier within the context of a larger flowchart, we don't necessarily have to draw the flowchart for the inside program. Rather, we might represent the entire program as a flattened hexagon. If a box indicates a decision, we use a diamond shape. A five-sided box portrays a part of the program that changes itself. A small circle identifies a processing junction point. Such a point in the program can go to several places. A small five-sided box, which has the shape of the home plate on a baseball field, shows where one page of a flowchart connects to the next, if the entire flowchart has more than one page. The intermediate junction and off-page connection points are labeled with numbers and letters to let readers know that all like symbols with the same character inside are meant to be connected together. Arrows indicate the direction of the flow.

Process paths

Returning to the flowchart for duplicating punched cards (Fig. 2-7), suppose that you want to change the card-punching program so that the computer skips blank cards and duplicates only those cards with some holes in them. Because the computer must make a decision about each card, you'll need to include a decision block in the flowchart. Figure 2-8 shows the result.

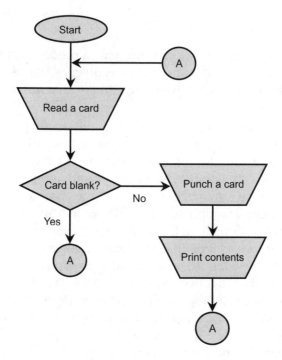

FIG. 2-8. *Example of a flowchart that includes a decision block (the diamond).*
The circles labeled A all represent a single junction point through which data flows
in the directions shown by the arrows.

Follow the flow

Except for the decision block, Fig. 2-8 shows the same process
as Fig. 2-7 does. The program begins in the "Start" oval at the top
and then goes to the block marked "Read a card." From there, the
program moves on to the decision block labeled "Card blank?"
If the answer is "Yes," the program proceeds to the connection
circle marked "A" and back to the top to read the next card. If the
answer is "No" (the card has holes in it), the program instructs the
hardware (the physical components of the computer) to punch
a duplicate card and print its contents. Then the program goes to
another circle marked "A" and back to the starting point.

Tip
Figure 2-8 is a simple flowchart, showing a process that uses only input and output devices and that does no calculations. Most programs and flowcharts involve more complicated processes.

The field of microcomputers uses many different types of diagrams that deal mostly with *software* (the operating systems and programs) rather than hardware (the physical components). From a purely electronic standpoint, functional diagrams abound and are usually more numerous than the schematic diagrams in the computer world. From an understanding standpoint, block diagrams can serve to display machine functions in general, but hardware maintenance and repair procedures require well-defined schematic drawings. Computers take advantage of the latest state-of-the-art developments in electronic components and are relatively simple from this standpoint, especially when you consider all they can do. However, from a pure electronics standpoint and as far as schematic diagrams are concerned, computers are highly complex; it would take many pages of schematics to represent even the most rudimentary computer.

Summary

Block diagramming can help you understand the general functioning of electronic circuits. Block diagrams are easy to draw, usually requiring only a marking instrument, some paper, and a straightedge (or a vector graphics computer program and a little bit of training on it). Schematic diagrams, in contrast, need more tools and can, in some cases, take many hours to render in a form that people can easily read and interpret.

3

Component symbols

On a road map, symbols illustrate towns, cities, secondary roads, primary roads, airports, railroad tracks, and geographical landmarks. The same rule applies to schematic drawings; symbols indicate conductors, resistors, capacitors, solid-state components, and other electronic parts. Every time a new component comes out, a new schematic symbol is derived for it. Often, a new type of component is a modification of one that already exists, so the new schematic symbol ends up as a modification of the symbol for the preexisting component.

Tip
In this chapter, you'll find most of the schematic component symbols commonly used in electricity and electronics. Appendix A contains a more complete listing in alphabetic, tabular form. You'll find it in the back of this book.

Resistors

Resistors are among the most simple electronic components. As the term implies, they resist the flow of electrical current. The value or "size" of a resistor is measured in units called *ohms*; typical real-world resistors are rated from about one ohm up to millions of ohms. Less commonly, you'll encounter resistors with values of less than an ohm, or hundreds of millions (or even billions) of ohms.

FIG. 3-1. *Standard schematic symbol for a fixed-value resistor.*

Regardless of the ohmic value, all fixed-value resistors are schematically indicated by the symbol shown in Fig. 3-1. This is the most universally accepted symbol for a resistor. The two horizontal lines indicate the leads or conductors that exit from both ends of the physical component. (Sometimes the resistor contacts are not wire leads but more substantial metal terminals.) Figure 3-2 shows a "transparent" functional drawing of a *carbon-composition* fixed resistor with leads on both ends. Figure 3-3 shows pictorial drawings of two other types of resistors. Any resistor of the sort shown pictorially in Fig. 3-2 or Fig. 3-3 is indicated schematically by the symbol in Fig. 3-1.

A variable resistor has the ability to change ohmic value by means of a slide or rotary tap that can be moved along the resistive element. The variable resistor is usually set to one value, and it remains at this point until manually changed. The electronic circuit, therefore, "sees" the component as a fixed resistor. However, when a variable resistor is required for the proper functioning of a specific circuit, it is necessary to indicate to any person who might build it from a schematic drawing that the resistor is actually a variable type. Figure 3-4 shows the schematic symbol for a variable resistor with two leads. Other types of variable resistors exist, and they have three leads (two end leads and a tap). Figure 3-5 shows two examples of schematic symbols for a three-terminal variable resistor, known as a *potentiometer* or a *rheostat* depending on the method of construction. Notice that

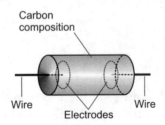

FIG. 3-2. *Pictorial illustration showing the anatomy of a carbon-composition resistor.*

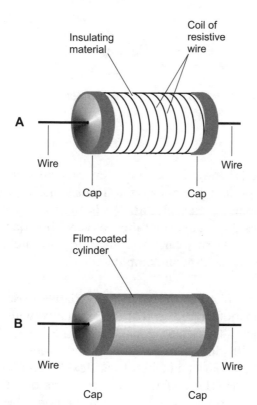

FIG. 3-3. *Pictorial illustrations showing the anatomy of a wirewound resistor (at A) and a film type resistor (at B).*

both examples use the standard resistor configuration, and indicate that it's a variable type by means of an arrow symbol pointing to the zig-zag part.

Did you know?

Rheostats are in effect the same as potentiometers, but mechanically they differ. A rheostat contains a wirewound resistance element, while a potentiometer is normally of the carbon-composition type. Therefore, a rheostat's value varies in small increments or steps, while a potentiometer's value can be adjusted over a continuous range.

FIG. 3-4. *Schematic symbol for a two-terminal variable resistor.*

Tip

In schematic drawings, an arrow often indicates variable properties of a component, but not always! Transistors, diodes, and some other solid-state devices have arrows in their schematic symbols. These arrows don't have anything to do with variable or adjustable properties. Arrows can also sometimes indicate the direction of current or signal flow in complex circuits.

Figure 3-6 is a pictorial drawing of a variable resistor of the wire-wound type, manufactured so that the resistance wire is exposed. A sliding metallic collar, which goes around the body of the resistor, can be adjusted to intercept different points along the coil of resistance wire. The collar is attached by a flexible conductor to one of the two end leads. The collar, therefore, shorts out more or less of the coil turns, depending on where it rests along the length of the coil. As the collar moves toward the opposite resistor lead, the ohmic value of the component decreases.

Figure 3-7 shows a functional drawing of a rotary potentiometer (at A), along with the schematic symbol (at B). The symbol looks like

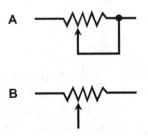

FIG. 3-5. *Alternate symbols for variable resistors, also known as potentiometers or rheostats (depending on the physical construction method). The device at A connects one end to the tap; and the device at B uses a three-terminal arrangement.*

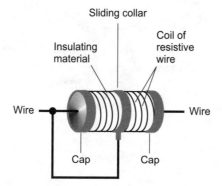

FIG. 3-6. *Pictorial drawing of a wirewound variable resistor.*

the variable resistor equivalent, but has three discrete contact points. Using the potentiometer control, the portion of the circuit that comes off the arrow lead can be varied in resistance to two circuit points, each connected to the two remaining control leads. Figure 3-8 shows a pictorial drawing of a typical potentiometer.

The variable resistor shown pictorially in Fig. 3-6 can be changed into a rheostat by severing the connection between the collar and

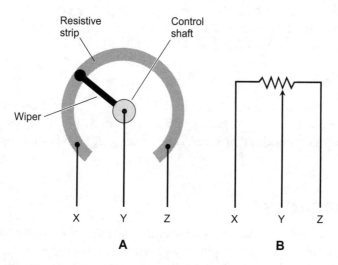

FIG. 3-7. *Simplified functional drawing of a rotary potentiometer (A) and its schematic symbol with corresponding connections (B).*

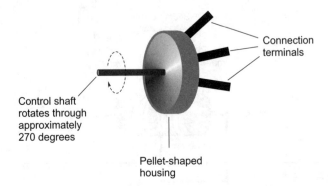

Connection
terminals

Control shaft
rotates through
approximately
270 degrees

Pellet-shaped
housing

FIG. 3-8. *Pictorial drawing of a full-size potentiometer, suitable for mounting on the front panel of an electronic device such as a radio receiver.*

the end. Now, the collar can be used as the third or variable contact. Likewise, a rheostat or potentiometer can be turned into a two-lead variable resistor by shorting out the variable contact point with the lead on either end.

The schematic symbol for a resistor, all by itself, tells us nothing about the ohmic value, or anything else about the component such as its power rating or physical construction, either. Various specifications for the component can be written alongside the resistor symbol, but these details might also appear in a separate components table and referenced by an alphabetic/numeric designation printed next to the schematic symbol (such as R1, R2, R3, and so on).

Tip
You can usually determine the ohmic value of a fixed resistor by looking at the colored bands or zones on it. Appendix B lists the resistor color codes that specify the ohmic values of fixed resistors.

Capacitors

Capacitors are electronic components that have the ability to block direct current (DC), while passing alternating current (AC). They also store electrical energy. The basic unit of capacitance is the *farad* (symbolized F). The farad is a huge electrical quantity, and most

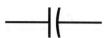

FIG. 3-9. *Standard symbol for a fixed capacitor. The curved line represents the plate (or set of plates) electrically closer to ground.*

real-world components are, therefore, rated in tiny fractions of a farad—*microfarads* or *picofarads*. A microfarad (symbolized µF) equals a millionth of a farad (0.000001 F), and a picofarad (symbolized pF) equals a millionth of a microfarad (0.000001 µF) or a trillionth of a farad (0.000000000001 F).

Figure 3-9 shows the common schematic symbol for a fixed capacitor. On occasion, you might see alternative symbols, such as those in Fig. 3-10. Many different types of capacitors exist. Some are *nonpolarized* devices, meaning that you can connect them in either direction and it doesn't make any difference. Others are *polarized*, having a positive and a negative terminal, and you must take care to connect them so that any DC voltage that happens to appear across them has the correct polarity. Most types of capacitors contain only two leads, although every now and then, you'll come across one with three or more leads.

The basic capacitor symbol consists of a vertical line followed by a space and then a parenthesis-like symbol. Horizontal lines connect to the centers of the vertical line and the parenthesis to indicate the component leads. The parenthesis side of a capacitor indicates the lead that should go to electrical ground, or to the circuit point more nearly connected to electrical ground. Unless the symbol includes a polarity sign, it indicates a nonpolarized capacitor, which might be made from metal plates surrounding ceramic, mica, glass, paper, or other solid nonconducting material (and, in some cases, air or a vacuum). The material designation indicates the insulation, technically known as a *dielectric*, that separates the two major parts of the component.

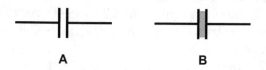

A   **B**

FIG. 3-10. *Alternate symbols for fixed capacitors. At A, air dielectric; at B, solid dielectric.*

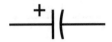

FIG. 3-11. *Schematic symbol for a polarized capacitor. The side with the plus sign (+) should carry a positive voltage relative to the other side.*

Physically, a typical fixed-value capacitor comprises two tiny sheets of conductive material close to each other but kept electrically separated by the dielectric layer.

Figure 3-11 shows the schematic symbol for a polarized or electrolytic capacitor. Notice that this symbol is the same as the one for the nonpolarized component, but a plus (+) sign has been added to one side. This sign indicates that the positive terminal of the component goes to the external circuitry. Occasionally, a negative (−) symbol will also appear on the opposite side. When you see the plus sign, you know that the component is polarized, and therefore, that you must connect it to the remainder of the circuit in observance of the proper polarity. That means the positive capacitor electrode must go to the more positive DC voltage point in the circuit, and the other electrode must go to the more negative DC voltage point in the circuit.

Tip
Polarized capacitors have external markings that tell you the polarity. Some have a plus sign, and some have a minus sign, and a few have both. Often, you'll need a magnifying glass to resolve the symbols, so beware: You should never connect a polarized capacitor the wrong way around!

All the capacitors that we've seen so far have a fixed design. In other words, the components specified have no provision for changing the capacitance value, which is determined at the time of manufacture. Some capacitors, however, do have the ability to change value. These components are generally called *variable capacitors*, although some specialized types are known as *trimmer capacitors* or *padder capacitors*.

Figure 3-12 shows the most common symbol for a variable capacitor. An arrowed line reveals the variable property; it runs diagonally through a fixed capacitor symbol. Figure 3-13 shows two alternative

FIG. 3-12. *Standard symbol for a variable capacitor. The curved line represents the rotor, and the straight line represents the stator.*

ways of indicating this same component. Most of the time, the symbol shown in Fig. 3-12 will indicate a variable capacitance, regardless of the physical construction details.

An *air variable capacitor* (one with an air dielectric) can tune many types of radio-frequency (RF) equipment including antenna matching networks, transmitter output circuits, and old-fashioned radios. A typical air variable has many interlaced plates, with the plates connected together alternately to form two distinct contact points. The set of plates that you can rotate is called the *rotor*; the set of plates that remains stationary is called the *stator*. All variable capacitors are nonpolarized components, meaning that the external DC voltage you connect to them can go either way and it doesn't make any difference.

Tip

In most air variables, the rotor should go to electrical ground. The rotor connects physically to the shaft that you turn. By grounding that shaft along with the rotor, you minimize *external capacitance* effects so that if you touch the shaft, the addition of your body into the system doesn't upset the performance of the circuit. In addition, you're protected against the risk of electric shock when the shaft that you touch goes directly to ground!

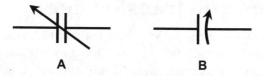

A   **B**

FIG. 3-13. *Alternate symbols for variable capacitors. At A, the stator is not distinguished from the rotor; at B, the rotor appears as a curved line with an arrow.*

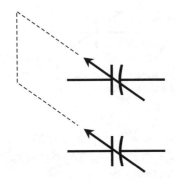

FIG. 3-14. *Schematic symbol for two variable capacitors ganged together.*

Sometimes, two separate variable capacitors are connected together or *ganged* in a circuit. In a ganged arrangement, two or more units are used to control two or more electronic circuits, but both components are varied simultaneously by tying the rotors of the two units together. Figure 3-14 shows the schematic symbol for two variable capacitors ganged together. The minimum and maximum capacitance values of the two components might be the same, but they don't have to be the same. They will, however, always track together. In a ganged system, when one of the capacitors increases in value, the others all increase as well.

As is the case with most electronic components, the schematic symbol for the capacitor serves only to identify it and to show whether it is fixed or variable, and if fixed, whether or not it is polarized. The component value might be written alongside the schematic symbol, or the component might be given a letter and number designation (for example, C1, C2, C3, and so on) for reference to a components list or table that goes along with the diagram.

Inductors and transformers

A basic *inductor* comprises a length of wire that is coiled up in order to introduce *inductance* into a circuit. Inductance is the property that opposes change in existing current; it acts in practice only while current increases or decreases. Coils or inductors can range in physical size from microscopic to gigantic, depending upon the inductance

FIG. 3-15. *Standard symbol for an air-wound (or air-core) inductor.*

value of the component, and on the amount of current that it can handle.

The basic unit of inductance is the *henry* (symbolized H), a large electrical quantity. Most practical inductors are rated in *millihenrys* (symbolized mH), where 1 mH = 0.001 H, or in *microhenrys* (μH), where 1 μH = 0.001 mH = 0.000001 H. Occasionally, you'll see an inductor whose value is specified in *nanohenrys* (nH), where 1 nH = 0.001 μH = 0.000000001 H.

Figure 3-15 shows the basic schematic symbol for an *air-core inductor*. The two leads are designated by straight lines that merge into the coiled part. An air-core coil has nothing inside the windings that can affect the inductance. Some air-core coils are wound from stiff wire and support themselves mechanically, and their cores do, in fact, comprise nothing but air. In most cases, however, a nonconductive and noninductive form made out of plastic, mica, or ceramic material serves as a support for the coil turns, keeping them in place and enhancing the physical ruggedness of the component.

Did you know?

In some old radio receivers, you'll find air-core inductors wound around small waxed cardboard cylinders resembling short lengths of drinking straw. Some hobbyists even use waxed wooden dowels to support "air-core" coils!

Figure 3-16 shows the schematic symbol for a tapped air-core inductor; in this case, the coil has two tap points along its length.

FIG. 3-16. *Schematic symbol for an air-core inductor with two taps.*

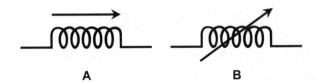

FIG. 3-17. *Schematic symbols for a variable air-core inductor. At A, arrow above coil symbol; at B, arrow passing through coil symbol.*

Whereas the fixed coil had only two leads, a tapped coil has three or more. When a coil is tapped, separate conductors are attached to one or more of the turns for intermediate connection. Maximum inductance is obtained from connecting the end leads to the external circuitry. A tapped arrangement allows for the selection of an input or output point that offers lower inductance than the full coil does.

As an alternative to taps, a coil might have a sliding contact that can be advanced along the entire length of the windings. This sliding contact allows adjustment of the inductance value, rather than having a select fixed point with the tapping arrangement. A variable coil can be indicated by either of the symbols shown in Fig. 3-17. The arrow indicates that the component can be adjusted from a maximum inductance value to a minimum inductance value.

Figure 3-18 shows symbols for a fixed air-core coil (at A), a tapped air-core coil (at B), and an adjustable air-core coil (at C).

An inductor meant for low-frequency applications can consist of a coiled wire wound around a solid or laminated (layered) iron core. Here, the iron, which constitutes a *ferromagnetic material*,

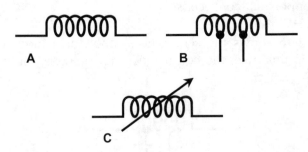

FIG. 3-18. *Schematic symbols for fixed (A), tapped (B), and adjustable (C) air-core inductors.*

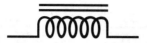

FIG. 3-19. *Schematic symbol for an inductor with a solid or laminated iron core.*

replaces the previous empty or air core. For example, a 60-Hz *choke*, intended for use in power-supply filters, will usually contain a single coil wound around a circular iron form. The ferromagnetic material greatly increases the *magnetic flux density* inside the coil windings, thereby increasing the inductance by a factor of many hundreds, or even thousands, of times compared with the inductance of an air-core coil having the same physical dimensions.

Figure 3-19 shows the schematic symbol for an iron-core inductor. Notice that it is the basic fixed coil discussed earlier, along with two close-spaced straight lines that run for its entire length. Sometimes the iron-core inductor is drawn as shown in Fig. 3-20, with the straight lines inside the coil turns in the symbol. (This is not the approved method of indicating an iron-core inductor, but you'll still see it now and then.) Some iron-core inductors contain taps for sampling different inductance values, and some might even be adjustable. The equivalent schematic symbols for these types of components appear in Fig. 3-21.

At higher frequencies, solid-iron and laminated-iron cores aren't efficient enough to function in inductors. Engineers would say that they have too much *loss*. At frequencies above a few kilohertz (kHz), a special core is needed if you want to increase the inductance over what you can get with *nonferromagnetic* core materials, such as air, plastic, ceramic, or wood. The most common substance for this purpose consists of iron material that has been shattered into myriad tiny fragments, each of which has a layer of insulation applied to it. After the fragmentation and insulation process has been completed, the particles are compressed to form a physically solid sample called

FIG. 3-20. *Alternate symbol for an inductor with a solid or laminated iron core.*

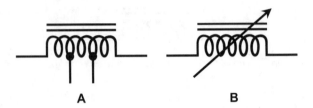

FIG. 3-21. *Symbols for a tapped coil (A) and an adjustable coil (B) with solid- or laminated-iron cores.*

a *powdered-iron core.* Figure 3-22 shows schematic symbols for powdered-iron-core inductors.

Tip
The symbols for powdered-iron-core inductors are nearly identical to those for solid- or laminated-iron-core inductors, except that the straight lines are broken up instead of solid. These types of components, like all other types of inductors, can be tapped or continuously variable.

A *transformer* is made up of multiple inductors with the coil turns interspersed or wound around different parts of a single core. Figure 3-23 shows the symbol for a basic air-core transformer. It looks like

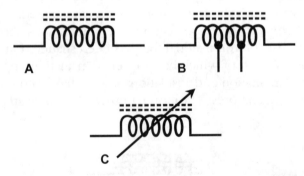

FIG. 3-22. *Schematic symbols for fixed (A), tapped (B), and adjustable (C) inductors with powdered-iron cores.*

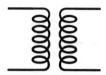

FIG. 3-23. *Schematic symbol for a transformer with an air core.*

two air-core coils drawn back-to-back. A transformer has the ability to transfer AC energy from one circuit to another at the same frequency. Because transformers are made by combining inductors, the schematic symbols are similar. Figure 3-24 shows some transformers that contain iron cores. The ones at A and B have solid or laminated cores; the ones at C and D have powdered cores.

Switches

A *switch* is a device, mechanical or electrical, that completes or breaks the path of current. Additionally, a switch can be used to allow

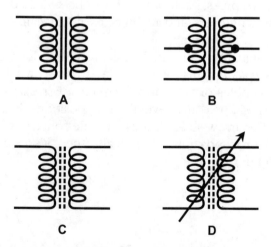

FIG. 3-24. *At A, a transformer with a solid- or laminated-iron core. At B, a transformer with a solid- or laminated-iron core and tapped windings. At C, a transformer with a powdered-iron core. At D, an adjustable transformer with a powdered-iron core.*

FIG. 3-25. *Schematic symbol for an SPST switch.*

current to pass through different circuit elements. Figure 3-25 shows the schematic symbol for a single-pole/single-throw (SPST) switch. This component can make or break a contact at only one point in a circuit; it's a two-position device (on-off or make-break).

Figure 3-26 shows a different type of switch, designated as a single-pole/double-throw (SPDT) component. Symbolically, the *pole* coincides with the point of contact at the base of the arrowed line. A *throw* is the contact point to which the arrow can point. The SPDT switch contains one pole contact and two throw positions; the input to the pole can be switched to either the upper or lower circuit point.

Some switches contain two or more poles. Figure 3-27A shows the symbol for a double-pole/single-throw (DPST) switch, while Fig. 3-27B shows the symbol for a double-pole/double-throw (DPDT) switch. Some switches have even more elements. The one shown in Fig. 3-28 has five poles, each of which can be switched to two separate positions. Engineers and technicians might call it a five-pole/double-throw (5PDT) arrangement.

This last designation can actually be covered under the heading of *multicontact switches*. This category takes in most switches that have more than two poles or two throw positions. For example, a rotary switch has a single pole and several throw positions; Fig. 3-29 shows an example. The arrow still indicates the pole contact. In this case the switch has 10 throw positions. Technically, then, it's a single-pole/10-throw (SP10T) device!

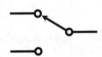

FIG. 3-26. *Schematic symbol for an SPDT switch.*

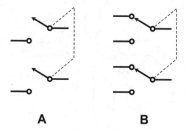

A   **B**

FIG. 3-27. *At A, symbol for a DPST switch. At B, symbol for a DPDT switch.*

Occasionally, you'll encounter sets of rotary switches ganged together, much like two or more variable capacitors can be made to rotate in sync with one another. Figure 3-30 shows the schematic symbol for an arrangement that uses two rotary switches. The dashed line tells us that the two switches are ganged. The two arrowed lines, which indicate the throw positions, go around "in sync" with each other. So, for example, when the left-hand switch (or pole number 1) rests at throw number 3 (as is the case here), the right-hand switch (or pole number 2) also rests at throw number 3.

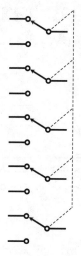

FIG. 3-28. *Schematic symbol for a five-pole double-throw (5PDT) switch.*

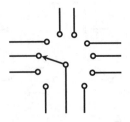

FIG. 3-29. *Schematic symbol for a rotary (or wafer) switch. This one has a single pole and 10 throws (SP10T).*

In each case, every switch contact point (pole or throw) is represented by a tiny circle. The variable element or pole is indicated by an arrow. The symbols shown here are all standard. You'll seldom see any significant variations.

Did you know?

Some amateur radio operators use a special switch called a *Morse code key.* This old-fashioned device, also called a *hand key* or a *straight key*, makes or breaks a circuit for the purpose of sending Morse code manually. It's an SPST switch with a lever and a spring, causing the device to return to the off position when the operator lets go of the lever. Figure 3-31 shows its schematic symbol.

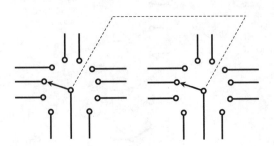

FIG. 3-30. *Schematic symbol for two rotary switches ganged together. This one has two poles and 10 throws (2P10T).*

FIG. 3-31. *Schematic symbol for a Morse code key (hand key or straight key).*

Conductors and cables

Throughout this discussion, a straight line has always indicated a conductor, but most circuits contain a large number of conductors. When you draw a diagram of a complicated circuit or system, you'll often find it necessary to have lines cross over each other, whether the represented wires actually make contact in the physical system or not.

Figure 3-32 shows two conductors that must cross each other in a diagram, but that are not connected to each other in the physical circuit (at least not at the point where they cross in the schematic). This diagram geometry does not imply that when you build the circuit, the conductors must physically cross over each other at that exact place. It simply means that in order to make the schematic drawing, you have to draw one conductor across another to reach various circuit points without introducing a whole lot of confusion and clutter, or resorting to three dimensions to make your drawing.

> **Aha!**
>
> A real-world circuit exists in three-dimensional (3D) space, but when you want to diagram it, you must do it on a two-dimensional (2D) surface. To carry off that feat, you must learn a few tricks to make sure that your readers see things right!

FIG. 3-32. *Schematic symbol for conductors that cross paths but are not electrically connected.*

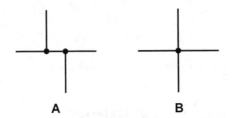

FIG. 3-33. *At A, preferred symbol for conductors that intersect and are electrically connected to each other. At B, alternate symbol for the same situation.*

Figure 3-33 shows two ways of portraying a point where two wires cross and they are electrically connected at that point. In the drawing at A, one of the conductors is "broken in two" so that it appears to contact the other one at two different points. This geometry makes it clear that the two conductors (the "divided" vertical one and the "solid" horizontal one) connect to each other electrically. Black dots indicate electrical connection. In the drawing at B, the two conductors cross (at right angles in this example), and a single black dot is drawn at the junction. This dot tells us that the conductors connect at this point. The method shown at B might look better at first glance, but the neatness comes along with a problem: Some readers might overlook the black dot and think that the two conductors are *not* meant to connect. The method at A makes that potential misinterpretation impossible.

Just as a reader might miss a black dot at a crossing point, as in Fig. 3-33B, another reader might see Fig. 3-32 and imagine a black dot when it isn't there! Then the reader will think the two wires connect when in fact they do not. This problem rarely occurs in well-engineered schematics where the draftsperson makes sure to use big black dots and good quality printing presses. However, in some older schematics you will see nonconnecting, crossed wires shown as in Fig. 3-34. One of the wires has a half loop that makes it look like it jumps over the other wire to avoid contact. That trick (which should never have gone out of style, in my opinion) gets rid of any doubt as to whether the wires electrically connect at the crossover point or not.

A *cable* consists of two or more conductors inside a single insulating jacket. In many cases, unshielded cables are not specifically

FIG. 3-34. *Archaic (but clear) representation of conductors that cross paths but are not electrically connected to each other.*

indicated in a schematic drawing, but appear as two or more lines that run parallel to indicate multiple conductors. Shielded cables require additional symbology along with the conductors. Figure 3-35 shows examples of shielded wire, often used to indicate the use of *coaxial cable* in an electronic circuit. Coaxial cable contains a single wire called the *center conductor* surrounded by a cylindrical, conduit-like conductive *shield.* An insulating layer, called the *dielectric,* keeps the two conductive elements isolated from each other. In most coaxial cables, the dielectric material consists of solid or foamed polyethylene.

Tip
Figure 3-36 shows a symbol for coaxial cable when the shield connects to a *chassis ground,* such as the metal plate on which an electronic circuit is constructed. The chassis ground might lead to an earth ground, but that's not always the case. In a truck, for example, no earth ground exists, so the chassis of the trucker's CB radio would go to the vehicle frame.

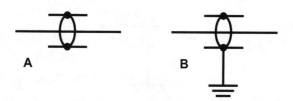

A   B

FIG. 3-35. *At A, symbol for a coaxial cable with an ungrounded shield. At B, symbol for a coaxial cable with an earth-grounded shield.*

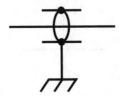

FIG. 3-36. *Symbol for a coaxial cable with a chassis-grounded shield.*

In some cables, a single shield surrounds two or more conductors. Figure 3-37 shows the schematic symbol for a two-conductor shielded cable. This symbol is identical to the one for coaxial cable, except that an extra inner conductor exists. If more than two inner conductors exist, then the number of straight, parallel lines going through the elliptical part of the symbol should equal the number of conductors. For example, if the cable in Fig. 3-37 contained five conductors, then five horizontal lines would run through the elliptical part of the symbol.

Diodes and transistors

Figure 3-38 shows the basic symbol for a *semiconductor diode*. In this symbol, an arrow and a vertical line indicate parts of the diode, and the horizontal lines to the left and right indicate the leads. The symbol in Fig. 3-38 portrays a *rectifier diode*. The arrowed part of the symbol corresponds to the diode's *anode*, and the short, straight line at the arrow's tip corresponds to the *cathode*. Under normal operating conditions, a rectifier diode conducts when the electrons move

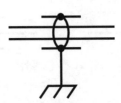

FIG. 3-37. *Symbol for a shielded two-conductor cable, in this case with a chassis ground for the shield.*

FIG. 3-38. *Symbol for a general-purpose semiconductor diode or rectifier.*

against the arrow, i.e., when the anode has a positive voltage with respect to the cathode.

Figure 3-39 shows the symbols for some specialized diode types. At A, we see a *varactor diode*, which can act as a variable capacitor when we apply an adjustable DC voltage to it. At B, we see a *Zener diode*, which can serve as a voltage regulator in a power supply. At C, we see a *Gunn diode*, which can act as an oscillator or amplifier at microwave radio frequencies.

A *silicon-controlled rectifier* (SCR) is, in effect, a semiconductor diode with an extra element and corresponding terminal. Its schematic symbol appears in Fig. 3-40. In the SCR representation, a circle often (but not always) surrounds the diode symbol, and the control element, called the *gate*, shows up as a diagonal line that runs outward from the tip of the arrow. In all cases, the lead that goes to the base of the arrow is the *anode* of the device, and the one connected to the short straight line at the arrow's tip is the *cathode*.

Figure 3-41 shows the schematic symbols for *bipolar transistors*. The PNP type is shown at A, followed by the NPN variety at B. The only distinction between the two is the direction of the arrow. In the PNP device, the arrow points into the straight line for the base electrode. In the NPN device, the arrow points outward from the base. Occasionally, the circle that surrounds the base, emitter, and

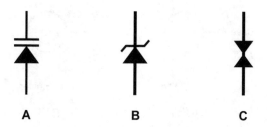

A   **B**   **C**

FIG. 3-39. *Symbols for a varactor diode (A), a Zener diode (B), and a Gunn diode (C).*

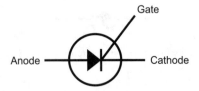

FIG. 3-40. *Symbol for a silicon-controlled rectifier (SCR).*

collector leads is omitted from the bipolar transistor symbol. Besides the bipolar variety, many other types of transistors exist. Figure 3-42 shows the symbols for four of these devices, as follows:

- At A, we see an *N-channel junction field-effect transistor* (JFET).
- At B, we see a *P-channel JFET.*
- At C, we see an *N-channel metal-oxide-semiconductor field-effect transistor* (MOSFET).
- At D, we see a *P-channel MOSFET.*

Tip

Transistors can be made from various types of semiconductor materials and metal-oxide compounds, but the schematic symbol, all by itself, tells us nothing about the elemental semiconductor material used in manufacture. The symbol merely indicates component functionality.

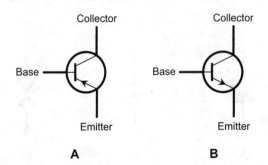

FIG. 3-41. *Symbols for a PNP bipolar transistor (A) and an NPN bipolar transistor (B).*

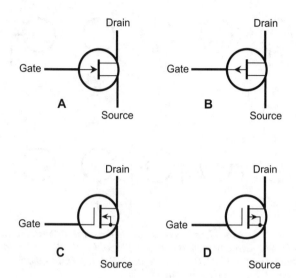

FIG. 3-42. *At A, symbol for an N-channel JFET. At B, symbol for a P-channel JFET. At C, symbol for an N-channel MOSFET. At D, symbol for a P-channel MOSFET.*

Electron tubes

Although vacuum tubes aren't used in electronics nearly as often as they were a few decades ago, many designs still exist that do employ them. When you want to create the symbol for a vacuum tube, you should start by drawing a fairly large circle, and then you should add the necessary symbols inside the circle to symbolize the type of tube involved. Figure 3-43 shows the schematic symbols for the various types of tube elements commonly used in schematic drawings.

Figure 3-44 shows the schematic symbol for a *diode vacuum tube.* This two-element device contains an *anode* (also called a *plate*) and a *cathode.* Just as with the semiconductor diode, the anode is normally positive with respect to the cathode when the device conducts current. The cathode emits electrons that travel through the vacuum to the anode. A hot-wire *filament,* something like a miniature low-wattage light bulb, heats the cathode to help drive electrons from it. In Fig. 3-44, the filament has been omitted for simplicity, a common practice in all vacuum tube symbology when the filament and cathode are physically separate, an arrangement known as an *indirectly heated cathode.*

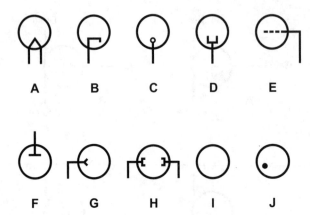

FIG. 3-43. *Symbols for tube elements and characteristics. A: Filament or directly heated cathode. B: Indirectly heated cathode. C: Cold cathode. D: Photocathode. E: Grid. F: Anode (plate). G: Deflection plate. H: Beam-forming plates. I: Envelope for vacuum tube. J: Envelope for gas-filled tube.*

Tip

All tube elements are surrounded by a circle, which represents the tube envelope. Occasionally, the circle is omitted from some tube symbols in schematic drawings, but that's not standard practice.

Figure 3-45 shows two versions of a *triode* vacuum tube, which consists of the same elements as the diode previously discussed, with the addition of a dashed line to indicate the grid. But there's another difference, too, in drawing A. Can you see it? Look closely

FIG. 3-44. *Schematic symbol for a diode vacuum tube with an indirectly heated cathode. Although a filament exists, it is often omitted to reduce clutter in symbols for tubes with indirectly heated cathodes.*

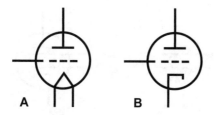

FIG. 3-45. *Symbols for a triode tube with a directly heated cathode (A) and an indirectly heated cathode (B).*

at the cathode. The tube at A has a *directly heated cathode,* in which the filament and the cathode are the very same physical object! We apply the negative cathode voltage directly to the filament wire; no separate cathode exists at all. In Fig. 3-45B, we see the symbol for a triode tube with an indirectly heated cathode. In this symbol, the filament is inside the cathode, which comprises a metal cylinder running along the central vertical axis of the tube.

Tetrode vacuum tubes have two grids. To represent one of them, we need an additional dashed line, as shown in the drawings of Fig. 3-46. In the tetrode, the upper grid, closer to the anode, is called the *screen.* Figure 3-47 shows symbols for the so-called *pentode* tube, which has three grids and a total of five elements. In the pentode, the second grid (going from the bottom up) is the screen, and the third grid (just underneath the plate) is called the *suppressor.* In both Figs. 3-46 and 3-47, the left-hand symbol (at A) portrays a device with a directly heated cathode, while the right-hand drawing (at B) shows a device with an indirectly heated cathode.

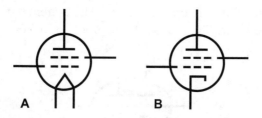

FIG. 3-46. *Symbols for a tetrode tube with a directly heated cathode (A) and an indirectly heated cathode (B).*

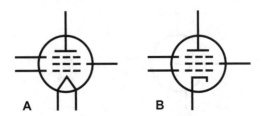

FIG. 3-47. *Symbols for a pentode tube with a directly heated cathode (A) and an indirectly heated cathode (B).*

Follow the flow

In all the vacuum tube symbols shown here, electrons normally flow from the bottom up. They come off the cathode, travel through the grid or grids (if any), and end up at the plate. Once in awhile you'll see a vacuum tube symbol lying on its side. In that sort of situation, you can simply remember that the electrons go from the cathode to the plate under normal operating conditions.

Some vacuum tubes consist of two separate, independent sets of electrodes housed in a single envelope. These components are called *dual tubes.* If the two sets of electrodes are identical, the entire component is called a *dual diode, dual triode, dual tetrode,* or *dual pentode.* Figure 3-48 shows the schematic symbol for a dual triode vacuum tube with indirectly heated cathodes.

In some older radio and television receivers, tubes with four or five grids were sometimes used. These tubes had six and seven elements, respectively, and were called *hexodes* and *heptodes.* These esoteric devices were used mainly for *mixing,* a process in which two RF signals having different frequencies are combined to get new signals at the sum and difference frequencies. The schematic symbol for a hexode is shown in Fig. 3-49A; the symbol for a heptode is shown

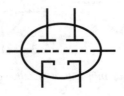

FIG. 3-48. *Schematic symbol for a dual triode tube.*

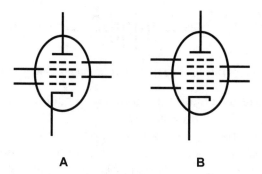

A   **B**

FIG. 3-49. *At A, symbol for a hexode tube. At B, symbol for a heptode tube, also known as a pentagrid converter.*

in Fig. 3-49B. Some engineers called the heptode tube a *pentagrid converter*. Both of these symbols show devices with indirectly heated cathodes.

> **Tip**
> You won't encounter hexodes and heptodes in modern electronics, but if you like to work with antique radios, you should get familiar with them. But take this warning: You'll probably have a difficult time finding a replacement component, should one of these relics go "soft" on you!

Cells and batteries

A cell or battery is often used as a power source for electronic circuits. Figure 3-50 shows the schematic symbol for a single *electrochemical cell*, such as the sort that you'll find in a flashlight. A single-cell component such as this usually has an output of approximately 1.5 V DC. *Electrochemical batteries* with higher voltage outputs comprise multiple cells connected in series (negative-to-positive in a chain or string);

FIG. 3-50. *Schematic symbol for a single electrochemical cell.*

FIG. 3-51. *Schematic symbol for a self-contained multicell electrochemical battery.*

the schematic representation for a multicell battery takes this design into account, as shown in Fig. 3-51.

The multicell battery symbol is simply a number of single-cell symbols placed end-to-end without any intervening lines. If a circuit calls for the use of three individual, discrete single-cell batteries in a series connection, you might draw three cell symbols in series with wire conductor symbols between them (Fig. 3-52). Alternatively, if multiple individual cells are set in a "battery holder" designed for direct series connection, you can use a battery symbol to portray the whole bunch.

Standard practice calls for polarity signs to go with the symbols for cells or batteries. Unfortunately, some draftspeople neglect this detail. Then when you see the schematic, you'll have to infer the polarity by scrutinizing the rest of the circuit.

Logic gates

All digital electronic devices employ switches that perform specific logical operations. These switches, called *logic gates,* can have anywhere from one to several inputs and (usually) a single output. Logic devices have two states, represented by the digits 0 and 1. The 0 digit is normally called "low" and the 1 digit is called "high."

- A *logical inverter,* also called a *NOT gate,* has one input and one output. It reverses, or inverts, the state of the input. If the input equals 1, then the output equals 0. If the input equals 0, then the output equals 1.
- An *OR gate* can have two or more inputs (although it usually has only two). If both, or all, of the inputs equal 0, then the

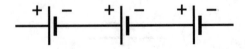

FIG. 3-52. *Symbol for three single electrochemical cells connected in series to form a battery.*

output equals 0. If any of the inputs equals 1, then the output equals 1. Mathematical logicians would tell us that such a gate performs an *inclusive-OR operation* because it "includes" the case where both variables are high.

- An *AND gate* can have two or more inputs (although it usually has only two). If both, or all, of the inputs equal 1, then the output equals 1. If any of the inputs equals 0, then the output equals 0.
- An OR gate can be followed by a NOT gate. This combination gives us a *NOT-OR gate,* more often called a *NOR gate.* If both, or all, of the inputs equal 0, then the output equals 1. If any of the inputs equals 1, then the output equals 0.
- An AND gate can be followed by a NOT gate. This combination gives us a *NOT-AND gate,* more often called a *NAND gate.* If both, or all, of the inputs equal 1, then the output equals 0. If any of the inputs equals 0, then the output equals 1.
- An e*xclusive OR gate,* also called an *XOR gate,* has two inputs and one output. If the two inputs have the same state (either both 1 or both 0), then the output equals 0. If the two inputs have different states, then the output equals 1. Mathematicians use the term *exclusive-OR operation* because it doesn't "include" the case where both variables are high.

Figure 3-53 illustrates the schematic symbols that engineers and technicians use to represent these gates in circuit diagrams.

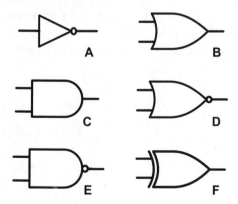

FIG. 3-53. *Symbols for a logical inverter or NOT gate (A), an OR gate (B), an AND gate (C), a NOR gate (D), a NAND gate (E), and an XOR gate (F).*

Summary

You'll encounter lots of symbols in electronics other than the common ones shown in this chapter. Appendix A is a comprehensive table of schematic symbols. In addition to the ones already discussed, you will see symbols for jacks and plugs, piezoelectric crystals, lamps, microphones, meters, antennas, and many other electronic components.

It might, at first thought, seem like a massive chore to memorize all of these symbols, but their usage and correct identification will come to you with practice and with time. The best way to begin the learning process is to read simple schematics and refer to Appendix A in this book whenever a symbol crops up that you can't identify. Within a few hours you'll be able to move on to more complex schematics, again looking up the unknown symbols. After a few weekends of practice, you should be thoroughly familiar with most electronic symbols used in schematic representations, so that when you see one in a diagram, you'll recognize it without having to think about it.

Schematic symbols are the fundamental elements of a communication scheme, like the symbols in mathematical expressions or architectural blueprints. Most schematic symbols in electronics are based on the structure of the components or devices they represent. Schematic symbols often appear in groups, each of which bears some relationship to the others. For example, you'll encounter many different types of transistors, but they're all represented in a similar fashion. Minor symbol changes portray variations in internal structure, but all can be easily identified as some type of transistor. The same rule applies to the symbols for diodes, resistors, capacitors, inductors, transformers, meters, lamps, and most other electronic components.

4

Simple circuits

When it comes to teaching people how to read and draw schematic diagrams, two schools of thought prevail. One school feels that you should learn to read diagrams before you learn to draw anything. The other school feels that you should learn to read diagrams as part of the process of drawing them. Both sides of this issue have their merits, so let's take advantage of them both! Certainly, you must start out by learning the basic component symbols. Then you should try to read as many schematic drawings as you can find. When this business starts to grow boring, you can draw your own simple diagrams. If you alternately read and draw schematics as the moods strike, you'll get comfortable with them before long! I recommend that you devote half of your study time to reading and the other half to drawing.

Getting started

This chapter deals with reading and drawing diagrams of some simple electronic circuits shown in pictorial and schematic form. Using this method, you can actually see the physical layout for a device, and then see how the schematic representation derives from it. Some commercial schematics are produced in this manner. However, in most instances, a circuit is designed schematically first, and then built and tested from the schematic. If a circuit is experimental, some bugs will exist in the prototype, necessitating various component deletions, substitutions, or modifications. When these changes are made to the test circuit, the results are noted, and the schematic is changed accordingly. In the end, the finished and corrected schematic is a product of design theory, actual testing, and modification.

Figure 4-1 shows a simple circuit that everyone has used at one time or another. Basically, it's a flashlight with the external case and the on/off switch removed. The device consists of a single electrochemical cell and an electric light bulb. This pictorial representation also shows the conductors, which attach to the light bulb and the battery. The conductors provide a current path between the battery and the light bulb.

Follow the flow

In the circuit of Fig. 4-1, electrons travel from the negative terminal of the cell through the bulb element and back to the positive terminal of the cell, "leapfrogging" from atom to atom in the metal wire and the bulb filament. That's how most electricians and engineers look at this situation. But some physicists will tell you that the current actually goes from the positive cell pole to the negative cell pole. That's called *theoretical current* or *conventional current*.

In order to draw a schematic diagram of the flashlight drawn in Fig. 4-1, you need to know three schematic symbols. These represent the electrochemical cell, the conductors, and the bulb, as shown in Fig. 4-2. Once you know the symbols, you can assemble them in a

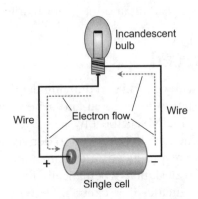

FIG. 4-1. *Pictorial drawing of a flashlight circuit using a single electrochemical cell (or single-cell battery), some wire, and an incandescent bulb.*

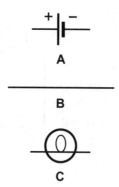

A

B

C

FIG. 4-2. *Schematic symbols for an electrochemical cell (A), an electrical conductor such as wire (B), and an incandescent bulb (C).*

logical manner based on the appearance of the circuit in the pictorial drawing.

Start by drawing the cell symbol. You can think of the cell as the heart of the circuit because it supplies all of the power for the device; it "pumps the electrons" through everything! Next comes the symbol for the light bulb, which you can draw at any point near the cell. Using this example, you should try to make the schematic symbols fall in line with the way the pictorial diagram appears. This layout places the light bulb above the cell.

Now that you've drawn the two major symbols, you can use the conductor symbols (plain, straight, solid, black lines) to hook them together. Notice that the pictorial drawing shows two conductors. Therefore, the schematic diagram also has two conductors. Figure 4-3 shows the completed schematic drawing, which is the symbolic equivalent of Fig. 4-1.

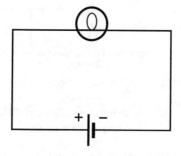

FIG. 4-3. *Schematic diagram of the single-cell flashlight.*

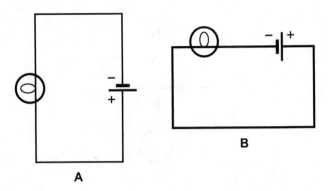

FIG. 4-4. *Alternative arrangements for the flashlight schematic. At A, cell and bulb "vertical;" at B, cell and bulb both on top.*

Figure 4-3 is by no means the only way that you can represent this simple circuit in schematic form. But any schematic representation will require the use of the same three basic symbols: cell, bulb, and conductors. The only changes that can occur involve the positioning of the component symbols on the page. Figure 4-4 shows two different alternatives for portraying the same circuit. All three of these diagrams (the one in Fig. 4-3 and the two in Fig. 4-4) are electrically equivalent, but they look somewhat different as a result of the relative positions of the components on the page.

Let's change this circuit a little bit, in order to gain proficiency in reading and writing schematics. Figure 4-5 shows the same basic flashlight circuit, but an additional cell and a switch have been added. This

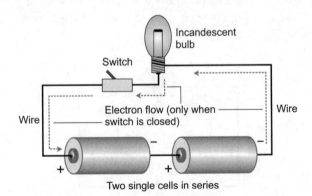

FIG. 4-5. *Pictorial drawing of a flashlight circuit using two single cells in series, some wire, a switch, and an incandescent bulb.*

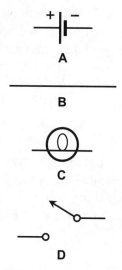

FIG. 4-6. *Symbols for components in the two-cell switched flashlight: Cell (A), wire (B), bulb (C), and switch (D).*

configuration is quite common for flashlights sold in the United States. By examining this pictorial drawing, you can see that any schematic representation will need symbols for the cells, the conductors, the light bulb, and the switch. Figure 4-6 shows the symbols that you'll need to produce an accurate and complete schematic drawing of this circuit. Again, you should draw the symbols in the same basic order as the components are wired in the circuit. Figure 4-7 shows the resulting schematic. Note that the two cell symbols are drawn separately, connected in series, with polarity markings provided for each one. In the series connection, the positive terminal of one cell goes to the negative

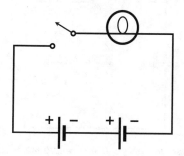

FIG. 4-7. *Schematic diagram of the two-cell switched flashlight.*

terminal of the other. The same two conductors are used from the cell terminals, but you need a third one to connect the switch to the light bulb, and you might also need a fourth one to connect the two cells together to form a battery (unless the cells rest directly against each other, a common state of affairs inside commercially manufactured flashlights). Figure 4-7 shows the switch in the off position.

Eureka!

Now you know what a common two-cell flashlight looks like when represented with schematic symbology. The next time that you switch one of those things on, you can imagine the switch symbol in Fig. 4-7 moving from the off (or open) position to the on (or closed) position.

Figure 4-8 is a pictorial representation of a device called a *field-strength meter*. Wireless communications engineers sometimes use this type of meter to see whether or not an RF electromagnetic (EM) field exists at a given location. You'll find this little circuit quite handy if you enjoy amateur radio, or if you need to locate the source of something that's causing RF interference. The circuit consists of an antenna, an RF diode, a microammeter (a sensitive current meter graduated in millionths of an ampere), and a coil. In order to draw this circuit

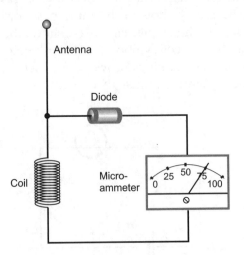

FIG. 4-8. *Pictorial representation of a field-strength meter circuit.*

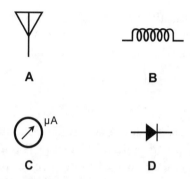

A   B

C   D

FIG. 4-9. *Schematic symbols for the components in the field-strength meter: Antenna (A), coil (B), microammeter (C), and diode (D).*

schematically, you need to know the symbols for an antenna, a coil, a microammeter, and a diode, all of which appear in Fig. 4-9. Using the same method as before, you can draw the schematic by connecting the symbols in the same geometric sequence as the components they represent appear in the circuit.

Figure 4-10 is a schematic of the field-strength meter shown pictorially in Fig. 4-8. This drawing involves nothing more than substitution of the schematic symbols for the pictorial symbols. As before, the parts need not be physically placed in the same positions as the schematic diagram suggests, but they must be interconnected precisely as indicated in the schematic. When you build a circuit from a schematic diagram that you trust, you should double-check and triple-check your actual component interconnections to make sure that they agree with the schematic. If you try to build the circuit shown in Fig. 4-10

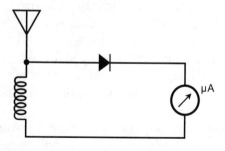

FIG. 4-10. *Schematic diagram of the field-strength meter.*

and make a mistake in the wiring connections, you cannot expect it to work. In more sophisticated devices and systems, wiring mistakes can cause component damage and, once in awhile, give rise to dangerous situations!

Follow the flow

In the circuit of Figs. 4-8 and 4-10, an EM field induces an RF current in the antenna and coil. That current is high-frequency AC. The diode rectifies the AC wave by "chopping off" either the positive half or the negative half of every cycle (depending on the diode's polarity) to produce pulsating DC like the output of a simple rectifier. The microammeter registers this current. As the strength of the EM field increases, the current increases, and the meter reading goes up.

Previously, we compared schematic drawings to road maps. A road map is supposed to indicate exactly what a motorist will experience in practice. The schematic drawing does the same thing. The highway lines that interconnect towns on a map correspond to the conductor lines that interconnect components in a schematic. A road map shows the route that a motorist takes while driving from town to town on the earth's surface, whereas a schematic diagram shows the route that a current or signal takes through the components in an electronic circuit.

Now let's look at something that's a little more complicated. Figure 4-11 is a schematic diagram of a *power supply* that produces pure, battery-like DC from utility AC. As you read this diagram from left to right, you'll see that a power plug goes to the transformer primary winding (the one to the left of the pair of vertical lines) through a fuse. At the top of the transformer secondary winding (the one on the right), a rectifier diode is connected in series. Following the diode, an electrolytic capacitor (note the polarity sign) is connected between the output of the rectifier and the bottom of the transformer secondary. A fixed resistor is connected in parallel with the capacitor. The DC output appears at the extreme right.

The physical size and weight of a real-world power supply, which you can build on the basis of Fig. 4-11, will depend on the voltage and current that you need to get from it. Because DC power

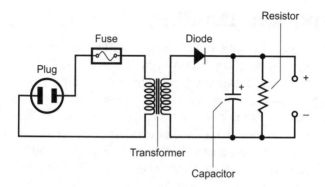

FIG. 4-11. *Schematic diagram of a simple DC power supply.*

supplies have polarized outputs, positive and negative signs indicate the output voltage polarity. Any power supply that uses a single diode, capacitor, and resistor will have this same basic configuration. Whether the output is 5 V at 1 A or 5000 V at 50 A, the schematic drawing will look the same. Figure 4-11 says nothing about how many volts or amperes the transformer, diode, capacitor, and resistor are meant to handle. You could add special features, such as a voltage regulator, overcurrent protector, voltmeter, or ammeter to your circuit and insert the symbols at the proper points in the schematic; but all half-wave DC power supplies are built around "cores" whose diagrams look like Fig. 4-11.

Follow the flow

In the circuit of Fig. 4-11, utility AC appears at the plug on the left. The AC travels through the fuse and flows in the transformer primary. In the secondary, AC also flows, but the voltage across the transformer secondary might be higher or lower than the voltage across the primary (depending on the transformer specifications). The diode allows current to flow only one way; in this case the electrons can go only from right to left (against the arrow). As a result, pulsating DC comes out of the diode. The capacitor gets rid of the pulsations, called *ripple*, on the DC output from the diode. The resistor discharges, or *bleeds*, the capacitor when you unplug the whole device from the utility outlet.

Component labeling

Figure 4-12 is a schematic of the same circuit as the one shown in Fig. 4-11, but in this case, each component has an alphabetic/numeric designation. These designators all refer to a components list at the bottom. Now you can see that this power supply uses a transformer with a primary winding rated at 125 V and a secondary winding that yields 12 V. The circuit has a diode rated at 50 peak inverse volts (PIV) and a forward current of 1 A; a 100-microfarad, 50-V capacitor; and a 10,000-ohm, 1-W carbon resistor. The fuse is rated at 0.5 A and 125 V.

The letters that identify each component are more or less standard. Notice that each letter is followed by the number 1. The designation T1, for instance, indicates that the component is a transformer (T) and that it's the first such component referenced. If this circuit had two transformers, then one of them would bear the label T1 and the other one would bear the label T2. The numbers reference the position or order on the components list; they serve no other purpose. The diode carries the reference designator D1, with D serving as the standard abbreviation for most diodes. Standardization is not universal, though! In some instances, the diode might bear the label SR1, where the letters SR stand for *silicon rectifier.* Some Zener diodes are labeled as ZD1, ZD2, and so on. This labeling makes little difference as long as you write the component designations next to the corre-

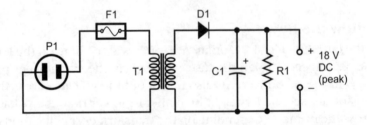

C1 - 100 microfarad electrolytic 50 V DC
D1 - 50 peak inverse volts at 1 ampere
F1 - 0.5 ampere at 125 volts
P1 - Male line plug
R1 - 10,000 ohm 1 watt carbon
T1 - 125 volt primary, 12 volt secondary, 1 ampere

FIG. 4-12. *Schematic diagram of the power supply with component designators and specifications.*

sponding symbols. If you replaced the designation D1 with SR1, your readers would still know that the abbreviation went with the symbol for the diode, as long as you made sure to put the abbreviation close enough to the symbol.

In the situation of Fig. 4-12, you don't have to include a number next to each component designation because only one of each component is used to make up the entire schematic! You could simply write P for the plug, F for the fuse, T for the transformer, D for the diode, C for the capacitor, and R for the resistor. Or, if you had confidence that your readers knew all the symbols, you could leave out designators altogether! Nevertheless, standard diagramming practice requires that you always include a letter and a number, even if only one of a certain component type exists in the whole circuit.

In complicated electronic systems, several hundred components of the same type (resistors, for example) might exist, many of which come from the same family. For instance, if you see the designation R101, then you know that the system contains at least 101 resistors. If you want to know the type and value of resistor R101, you will have to look up R101 in the components list to find its specifications.

Tip
You can use the schematic of Fig. 4-12 to build a power supply with a peak output of about 18 V DC. But before each component was referenced, the schematic had no practical use. Figure 4-11 illustrates a generic half-wave power supply, but that schematic doesn't give you any practical information other than the relative arrangement of the components.

Table 4-1 shows the standard letter designations for most types of electronic components that you'll encounter in schematic diagrams. Some of these designations can vary in real-world documentation, depending upon the idiosyncrasies of the person making the drawing or designing the circuit. You should find it easy to memorize (yes, that's right, memorize!) the information in Table 4-1 because most of the designations merely comprise the first letters of the component names. If the component has a complex name, such as *silicon-controlled rectifier*, the first letters from each of the three words is used, so you get SCR1. A resistor is designated by R, a capacitor by

Table 4-1
Letter designations for components that commonly appear
in schematic diagrams

Abbreviation	Full component name
ANT	Antenna
B	Battery
C	Capacitor
CB	Circuit board
D	Diode
EP	Earphone
F	Fuse
GND	Ground
I	Incandescent lamp
IC	Integrated circuit
J	Receptacle, jack, or terminal strip
K	Relay
L	Inductor
LED	Light-emitting diode
M	Meter
NE	Neon lamp
P	Plug
PC	Photocell
PH	Earphone
Q	Transistor
R	Resistor
RFC	Radio-frequency choke
RY	Relay
S	Switch or telegraph key
SCR	Silicon-controlled rectifier
SPK, SPKR	Speaker
SR	Silicon rectifier
T	Transformer
TP	Terminal or test point
U	Integrated circuit
V	Vacuum tube
Y	Quartz crystal
Z	Circuit assembly

C, a fuse by F, and so on. Conflicts do arise, of course. If you want to designate a relay, you need to use some letter other than R because R indicates a resistor! The same thing happens if you want to label a crystal; you can't use C because that letter refers to a capacitor. Look through Table 4-1 from time to time as you read and draw schematic diagrams, and eventually you'll absorb all the information in there.

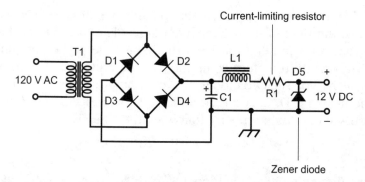

FIG. 4-13. *A power supply that uses full-wave bridge rectification (four rectifier diodes) and Zener-diode voltage regulation.*

The circuit of Fig. 4-13 has a full-wave bridge rectifier along with a better ripple filter than the simple capacitor used in the previous power supply. The inductor, L1, is a filter choke, which, along with capacitor C1, does an excellent job of "smoothing" out the DC so it resembles what comes from a 12-V battery (pure DC with no ripple).

Figure 4-14 shows a *voltage-doubler power supply.* The two capacitors, C1 and C2, charge up from the full transformer secondary output after the current goes through diodes D1 and D2. Because the two capacitors are connected in series, they act like two batteries in series, giving you twice the voltage. But there's a catch! A voltage doubler power supply works well only at low current levels. If you try to draw too much current from one of these power supplies, you'll "draw down" the capacitors and the voltage will decrease.

In Figs. 4-13 and 4-14, the letter designations are the same for each component type, but the numbers advance, one by one, up to the total number of units. So, for example, in Fig. 4-13 you see diodes

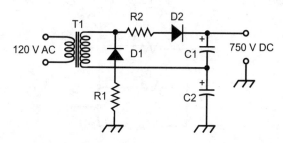

FIG. 4-14. *A voltage-doubler power supply.*

D1 through D5 because the circuit contains five diodes. (The Zener diode to the right of R1 has the letter D just like the rectifier diodes have, but you can tell it's a Zener diode because of the bent line in the symbol.) All the other components have only one of each type. In Fig. 4-14, you see two diodes, two capacitors, and two resistors, so the numbers for D, C, and R go up to 2. The transformer is all alone, so you see only the number 1 following the letter T.

Tip

Even though multiple components might all have the same value (820 ohms, for example, or 50 microfarads), they must nevertheless get separate *numerical* designations when two or more of them exist in a single circuit.

Schematics don't reveal every physical detail of a device, the way a photograph or detailed pictorial would do. Schematics depict schemes, that's all! The schematic diagram for a device allows engineers and technicians to make the correct electrical connections when putting it together, and to locate the various components when testing, adjusting, debugging, or troubleshooting it. If you find all this talk overly philosophical, maybe a real-world example will clear things up. Remember that solid lines in schematic drawings represent conductors. However, a conductor doesn't have to be a length of wire. It might be part of a component lead, or perhaps a *foil run* on a printed circuit board (the latter-day equivalent of a connecting wire). Whether or not a separate length of wire is needed to interconnect two components will depend on how close together those components are in the physical layout.

Examine the simple schematic of Fig. 4-15. The circuit contains three resistors, all of which go together in a parallel arrangement.

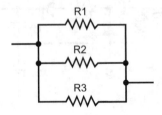

FIG. 4-15. *A simple circuit comprising three resistors in parallel.*

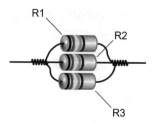

FIG. 4-16. *Pictorial diagram of three resistors in parallel with leads intertwined.*

Taking the schematic literally, a conductor connects the left-hand side of R1 to the left-hand side of R2. Another conductor goes between the left-hand side of R2 and the left-hand side of R3. Two other conductors connect the right-hand sides of the components. In practice, the connections might be made with wires attached to the resistor leads, but if the components are close enough together, the leads themselves can form the interconnections. Then Fig. 4-15 will represent the physical arrangement shown in Fig. 4-16.

Naturally, if you want to follow good engineering principles, you'll want to make all of your electronic circuits as compact (and dependable) as possible by using a minimum amount of point-to-point wiring and trying to make the component leads serve for interconnection purposes whenever you can. Of course, in the above example, if the three resistors had to go in different parts of the circuit separated by some physical distance, then you would need to use interconnecting conductors between them. However, as you design the physical layout of a circuit, you should try to minimize the overall length (that is, the total length) of all the interconnecting wires or foil runs combined.

Troubleshooting with schematics

Engineers and technicians use schematic diagrams to create electronic devices, but these diagrams can also prove invaluable for troubleshooting equipment when problems develop. Knowing how to read schematic diagrams, however, is not enough. You also need to know what tasks the various components actually perform, as well as how the diverse circuits work together in a complete system. No matter how proficient you might get at electronics troubleshooting, seemingly simple repair jobs can explode into major headaches without complete, accurate, and clear schematic representations of the hardware.

> ### Remember!
> Schematic diagrams clarify circuits. They present the circuit elements in a logical and easy-to-understand manner. They tell you very little, if anything, about the component layouts in actual devices.

When you build a circuit from a schematic drawing, the physical object rarely bears much physical resemblance to the schematic. It's impractical to build a complex electronic circuit by placing the components in the exact same geometrical relationship as they appear in the schematic. The diagrams purposely spread out the components on the page for easy reading. Schematic diagrams are two-dimensional, whereas real-world electronic components are three-dimensional. You need only to look inside of a major electronic device, such as a television set or computer, to realize the complexities that you'd face in troubleshooting a complex system without the help of a schematic diagram.

If you know a fair amount about electronic components and how they operate in various circuits, then you can use a schematic diagram to get a good idea (without any equipment testing) where a particular problem might occur. Then, by testing various circuit parameters at these critical points and comparing your findings with what the schematic diagram indicates should be present, you can make a quick assessment of the trouble. For example, if a schematic diagram shows a direct connection between two components in a circuit, and a check with an ohmmeter reveals a high resistance between the two, then you can assume that a conductor is broken or a contact has been shaken loose. If a schematic diagram shows only a capacitor between two components (with no other circuit routes around it) and a reading with your ohmmeter shows zero ohms or only a couple of ohms, you can assume that the capacitor has shorted out and you'll have to replace it.

Beginners to electronics troubleshooting and diagram reading sometimes assume that a professional can instantly isolate a problem to the component level by looking at the schematic. This idealized state of affairs might prevail for a few simple circuits, but in complex designs the situation grows a lot more involved. Often, the schematic diagram allows a technician to make educated guesses as to where or what the trouble might be, but an exhaustive diagnosis will nearly always require testing. A particular malfunction in an electronic device will not necessarily have a single, easy-to-identify cause. Often there are many possible causes, and the technician must whittle the situation down to a single cause by following a process of elimination.

Suppose that a circuit will not activate, and no voltage can be detected through testing at any contact point indicated by the schematic. Chances are good that no current is passing through the circuit at all. However, you don't know from this observation exactly what has caused the failure. Has one of the components in the power supply become defective? Has the line cord been accidentally pulled from the wall outlet? Has a conductor broken between the output of the power supply and the input to the electronic device? Has the fuse blown?

In a scenario of this sort, you will almost certainly want to consult the schematic diagram as you go through all of the standard test procedures. You might wish to find the contact point that serves as the power supply output, indicated on the schematic. If you test the voltage at this point and it appears normal, then you can assume that the problem lies somewhere further along in the circuit. The schematic diagram and the test instrument readings allow you to methodically search out and isolate the problem by starting at a point in the circuit where operation is normal and proceeding forward until you get to the point where the circuit shows some abnormality.

Continuing with the same example, if no output comes from the power supply, you know that you must search backward toward the trouble point. You will continue testing until you reach a point of normal operation and then proceed from there. Using a schematic diagram, you'll follow your progress and thereby narrow the problem area down to something between two points (the point farthest back from the output at which the problem exists, and the point furthest forward from the input where things test normal). Chances are good that this narrowing process will isolate the trouble to a single component or circuit connection.

Go back and look again at the flashlight circuit of Fig. 4-7. Although the schematic diagram does not say so, the two batteries in series should yield a DC potential of 3 V because a typical flashlight cell provides 1.5 V, and DC voltages add up in series connections. Some schematic diagrams provide voltage test points and maximum/minimum readings that you should expect, but this simple example doesn't.

Suppose that the flashlight has stopped working, and you decide to test the circuit with a *volt-ohm-milliammeter* (VOM), also called a *multimeter*, with the help of Fig. 4-7. First of all, you can measure the individual voltages across the cells. With the meter's positive probe placed at the positive cell terminal and the negative probe at the negative terminal, you should get a reading of 1.5 volts across each cell. If both read zero, then you know that both cells have lost all their electrical charge. If one cell reads normal and the other one reads zero, then in theory you should only have to replace the one that reads zero. (In practice, it's a good idea to replace entire sets of cells all at once, even if some of them still test okay). If both cells read normal, then you can test the voltage across the bulb. Here, you should expect a reading of 3 V under normal operation with the switch closed. If you do indeed observe 3 V here, then you can diagnose the problem by looking at the schematic. The bulb must have burned out! The schematic shows you that current must go through the light bulb if the bulb can conduct, so it must light up. If voltage is available at the base of the light bulb, then current will flow through the element unless it has opened up. But of course, if the bulb filament has broken apart, no current can flow through the bulb, so it won't light up. In fact, with a burned-out bulb, no current will flow anywhere at all in the circuit.

On the other hand, let's say that you get a normal reading at the batteries, but no reading whatsoever at the light bulb. Obviously, a break must exist in the circuit between these two circuit points. Three conductors are involved here: one between the negative terminal of the battery and one side of the bulb, another between the positive battery terminal and the switch, and another between the switch and the other side of the bulb. Obviously, one of the conductors has broken (or a contact has been lost where the conductor attaches to the battery), or maybe the switch is defective. While you keep an eye on the schematic, you can test for a defective switch by placing the negative meter probe on the negative battery terminal and the positive probe on the input to the switch. If you see a normal voltage reading, then the switch must be defective. If you still get no voltage reading, then one of the conductors has come loose or broken.

Admittedly, the scenario just described presents only a basic example of troubleshooting using a schematic diagram—almost as simple as things can get! But imagine that the flashlight circuit is highly complex, one you know nothing about. Then the schematic diagram becomes an invaluable aid and a necessary adjunct to the standard test procedures with the VOM. This same basic test procedure will be used over and over again when testing highly complex electronic circuits of a similar nature. In most instances, no matter how complicated the circuit design looks, it's actually a combination of many simple circuits. But if you have to do a comprehensive troubleshooting operation, you might have to test each and every one of those circuits individually.

A more complex circuit

Figure 4-17 shows a diagram for a somewhat more complicated, real-world electronic circuit presented in a form intended to assist a troubleshooting technician. The circuit has a single NPN bipolar transistor along with some resistors and capacitors. Note that *test points* (abbreviated TP) exist at three different locations: TP1 at the emitter of the transistor, TP2 at the base of the transistor, and TP3 at the collector of the transistor. If you need to troubleshoot this circuit (which happens to be a low-power amplifier of the sort you might find in a vintage radio receiver or hi-fi set), you'll connect your VOM between chassis ground and each one of these three points, in turn. You'll carefully note the meter readings and compare them with known normal values.

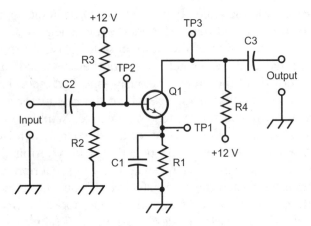

FIG. 4-17. *Schematic diagram of an amplifier circuit that includes component designators and three test point (TP) locations.*

Follow the flow

The circuit of Fig. 4-17 receives a weak input AC signal (such as the output of an ultrasonic pickup) and boosts it so that it can drive a device that consumes significant power (such as a switching circuit). The general signal flow goes from left to right. The original AC signal enters at the input terminals, passes through capacitor C2, and reaches the base (left-hand electrode) of transistor Q1. The base acts as an adjustable "current valve" that causes large current fluctuations through Q1 as the electrons flow from ground through R1 to the emitter (the bottom electrode), then onward to the collector (top electrode), and out through R4 to the positive power supply terminal. Capacitors C2 and C3 allow the AC signals to pass while blocking DC from the power supply so that the DC won't upset the operation of external circuits. Capacitor C1 keeps the transistor's emitter at a constant DC voltage while allowing the input signal to enter unimpeded. The resistors R2 and R3 have values carefully chosen to place precisely the right DC voltage, called *bias*, on the base of Q1, ensuring that the transistor will work as well as it possibly can in this application.

In many electronic circuits, actual voltages can deviate from design values by up to 20 percent; if this information is important, you'll usually find the error range at the bottom of the schematic drawing or in the accompanying literature. If the readings obtained are within this known error range (called the *component tolerance*), then you can tentatively assume that this part of the circuit is working properly. However, if the readings obtained are zero or well outside of the tolerance range, then you have pretty good reason to suspect a problem with the associated circuit portion, or possibly with other circuits that feed it.

Many schematic drawings that accompany electronic equipment, especially "projects" that you build from a kit containing individual components, include information that aids not only in troubleshooting, but also in the initial testing and alignment procedures that you must follow as soon as you've completed the physical assembly process. As a further aid, the literature might include pictorial diagrams that show you where each part belongs on the circuit board or chassis. That way, you can follow the circuit not only according to its electrical details, but along the physical pathways as they actually look.

According to standard schematic drawing practice, every component should bear a unique alphabetic/numeric label to designate it, as you see in Fig. 4-17. However, a few alternative labeling forms are also acceptable. Figure 4-18 shows the same circuit as the one in Fig. 4-17, but the parts list has been eliminated and the diagram contains no alphabetic/numeric designations. Instead, the compo-

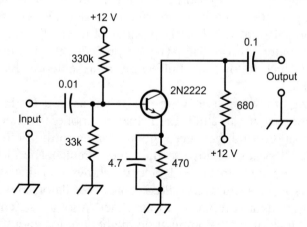

FIG. 4-18. *Schematic diagram of the amplifier circuit from Fig. 4-17, but with component values instead of sequential designators. All capacitances are in microfarads (μF). All resistances are in ohms (Ω); k = 1000.*

nents are identified only by their schematic symbols along with value designations or industry standard part designations. Using the example shown, we know that the transistor is a 2N2222 type, and that the resistors have values of 470, 33k (33,000), 330k (330,000), and 680 ohms. The input capacitor has a value of 0.01 microfarad (µF) and the output capacitor has a value of 0.1 µF. The emitter capacitor, which goes across the 470-ohm resistor, has a value of 4.7 µF.

Tip
In situations like the one in Fig. 4-18, you'll usually see a statement at the bottom of a schematic diagram that includes information about the units for the value designations. Such a statement might read "All capacitors are rated in microfarads (µF). All resistances are given in ohms (Ω), where k = 1000 and M = 1,000,000."

Schematic/block combinations

Once in awhile, you'll encounter a "hybrid" drawing that consists of a block diagram and a schematic diagram combined. Figure 4-19 shows an example of this approach, which works well when you want to highlight and explain a particular circuit, and to clarify its relationship to other circuits in a system. The detailed schematic portion of Fig. 4-19 shows a buffer amplifier of the sort you'll find in a radio transmitter. An oscillator (which precedes the buffer) and an amplifier (which follows the buffer) are represented as blocks with labels inside.

This diagram serves two purposes. First, as you read the schematic portion, you can study the actual component makeup of the buffer circuit. Second, you get a good idea as to the buffer's place in the overall system relative to the other circuits. The block/schematic representation in Fig. 4-19 clearly shows you that the buffer receives its input from a crystal-controlled oscillator, and also that the buffer sends its output to an amplifier. Another schematic diagram and block diagram combination might describe another portion of this same device, which you might recognize as a simple radio transmitter.

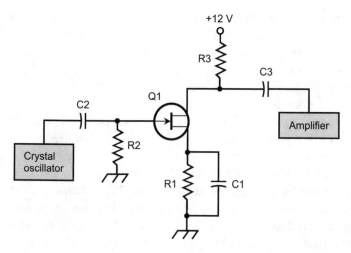

FIG. 4-19. *A combination block/schematic diagram of an oscillator, buffer, and amplifier system, showing the schematic details of the buffer portion.*

Figure 4-20 is a schematic/block "hybrid" diagram in which the oscillator is portrayed in schematic detail, but the buffer and the amplifier are represented only as blocks. This figure tells you that the oscillator output goes to the buffer, which then sends the signal along to the amplifier. The only new information obtained here is contained in

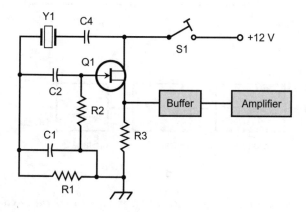

FIG. 4-20. *A block/schematic diagram of the system from Fig. 4-19, showing the schematic details of the oscillator portion.*

the schematic representation of the oscillator. If you look at Figs. 4-19 and 4-20 together, however, you can picture in your mind a diagram in which all of the system's detail is included up to the point where the signal enters the amplifier. The oscillator portion of Fig. 4-20 also tells you that the oscillator is designed to generate Morse code signals because it contains a telegraph key!

Now let's "shift gears" and recall the block diagram of the AC-to-DC converter that you saw all the way back in Fig. 2-1. Figure 4-21 shows the schematic representation of that device, along with a duplicate of Fig. 2-1 to save you the trouble of having to flip pages. As you compare the two diagrams, note that in the schematic (at A), all of the actual components are shown, rather than merely portraying the stages as labeled blocks (at B). Figure 4-22 is a more comprehensive diagram that shows how the block diagram of this same device relates to the schematic diagram, while revealing all of the details originally present in both diagrams.

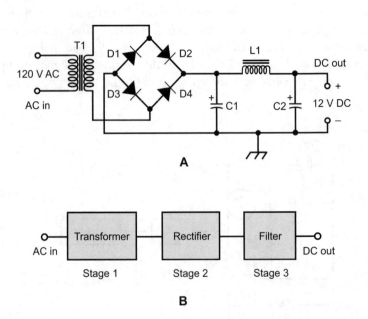

FIG. 4-21. *Schematic diagram of an AC-to-DC converter (at A), such as the one portrayed in the block diagram all the way back in Fig. 2-1 (reproduced here at B).*

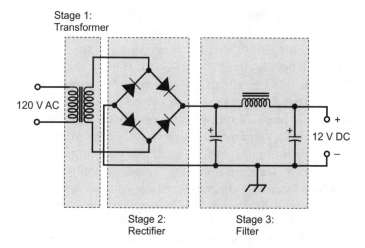

FIG. 4-22. *Here's how the block diagram of Fig. 2-1 relates to the schematic diagram of Fig. 4-21. (Component designators have been omitted to avoid clutter.)*

Follow the flow

To briefly explain the AC-to-DC converter circuit, more commonly called a power supply, the input electricity enters at the extreme left-hand end of all versions of the diagram (Figs. 4-21 and 4-22). The AC goes to the transformer to set up the proper ratio for conversion to DC. The four diodes in the diamond configuration constitute the rectifier, which converts the AC to pulsating DC, making use of both halves of the cycle. The final stage, the ripple filter, acts to smooth out the pulsations in the DC after the conversion. The "smoothed-out" or "pure" DC then goes along to the output terminals at the right-hand end, where it appears as a voltage just like the electricity that you'd find at the terminals of a 12-V battery.

Summary

This chapter has provided a brief sampling of simple electronic circuits, showing how you can read diagrams to learn how circuits are put together, or draw diagrams on the basis of known circuit details.

After a bit more training, you should be able to view a simple schematic and easily visualize what the finished circuit will look like, maybe even getting an idea of how you might go about arranging the components on a circuit board (even though the schematics or block diagrams don't explicitly tell you what to do in that respect).

You know that a roadway, as you drive along it in your car, does not look like the line that represents it on a map. You can visualize a secondary road or even a superhighway and do so whenever you see their symbols (although there's nothing like traveling along a road in both directions to learn all about it). The same principle holds true when reading schematic diagrams. When you see a bunch of schematic symbols in a diagram, you can visualize the component interconnections and perhaps make a few mental notes on the physical aspects of the circuit construction (although there's nothing like looking at, and using, a finished electronic device). Schematic symbology will become a new language for you—one in which you will eventually learn to think, just as you can think in terms of words, mathematical equations, musical scores, architectural blueprints, Morse code, or any other language.

5

Complex circuits

As you learn to read and draw schematic diagrams, you might get discouraged by the complexities that arise. You might think, "Oh sure, anyone can learn to read schematics of simple circuits, like those that have a transistor or two, a few capacitors, and a few resistors. But it must be impossible to decipher more complicated schematics without decades of experience." That's not true! You'll have to put some effort into the learning process, but you can always break a large, complicated system down into smaller, simpler circuits.

Identifying the building blocks

Even a circuit whose diagram looks overwhelming at first glance comprises multiple building blocks interconnected in an orderly, logical way. This principle never fails! Each building block represents a simple circuit. A device that contains six diodes, 10 inductors, 15 transistors, and dozens of resistors and capacitors might break down into 15 simple circuits, each containing a single transistor, a few resistors and capacitors, and a diode and inductor or two. If you look at the entire system schematic all at once, it's like trying to eat a jumbo hamburger in a single swallow. With the burger, and with the diagram as well, you face a far easier job if you assimilate the thing in little bites or pieces.

> **Tip**
> Nowadays, circuits of great complexity are etched onto semi-conductor wafers and housed in *integrated circuits* (ICs), also called *chips*, that come in packages that actually look like building blocks! As you examine a modern circuit board, you might imagine that the whole system was designed and put together from a block diagram, not a schematic! When you see the complete schematic, it will contain rectangles representing the chips, along with a great many lines connecting them. In most real-world circuits, those lines represent foil runs on a printed circuit board.

Figure 5-1 shows a "crystal radio" receiver built with an antenna, a tapped air-core inductor, a variable capacitor, an RF diode, and a fixed capacitor. The term "crystal" comes from the original construction of RF diodes as they emerged in the early 1900s. In order to get a one-way current gate to act as a signal *detector* (or *demodulator*), radio experimenters placed a piece of fine wire called a *cat's whisker* into contact with a piece of crystalline lead sulfide called *galena*. Today, semiconductor diodes do the same thing as those old "crystals" did, even though modern RF diodes don't look anything like crystals.

You can't call the circuit of Fig. 5-1 "complicated" by any stretch of your imagination, but it performs some sophisticated tricks nevertheless! Aside from the antenna (an external component such as a wire running out your window to a tree) and a sensitive earphone or headphone that you can connect at the output terminals in order to

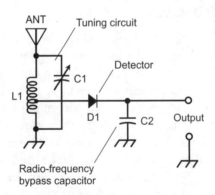

FIG. 5-1. *Schematic diagram of a "crystal radio" tuned circuit and detector stages. It produces an extremely weak audio signal.*

hear radio stations (however faintly), this circuit contains only four components: the coil, the diode, and two capacitors.

You can connect an amplifier to the output of the "crystal radio" in order to boost the audio volume to the point where it can drive a headset to a comfortable listening level. (That's the reason why Fig. 5-1 doesn't include a headset at the output.) To accomplish that feat, you'll need another circuit, along with a battery or other source of DC power to make the output signal strong enough.

Follow the flow

In the "crystal radio," RF energy from the antenna *resonates* (reverberates) in L1 and C1. The *resonant frequency* depends on the size of the coil and also on the setting of C1. That frequency determines which station you hear, if any. Diode D1 converts the signal to pulsating DC that contains both the RF and the audio from the original transmitter. Capacitor C2 shorts the RF part of the signal to ground, leaving only audio-frequency (AF) energy at the output. The position of the coil tap must be adjusted experimentally for maximum output.

Figure 5-2 shows another fairly simple schematic diagram: an audio amplifier that makes use of a single NPN bipolar transistor. In addition, this circuit has four resistors and three capacitors for a total of eight components. It also needs a source of DC power, such as a battery, which provides 12 V. This circuit can accept low-level audio signals (the output of a "crystal radio," for example) at the input terminals and boost the power to a level strong enough to make a nice loud sound come out of a headset. It might take an experienced engineer two or three minutes to draw the diagram, and a couple of hours to build and test the complete circuit, "tweaking" component values to get the best possible performance.

Because the circuit of Fig. 5-2 takes an extremely weak signal and boosts it to a reasonable (but not particularly powerful) level, it's sometimes called a *preamplifier*. If you want to drive a set of speakers so that the students in a classroom, or the audience in an auditorium, can hear the output from the "crystal radio," you'll need more amplification. That extra audio boost can be provided by one or more additional circuits called *power amplifiers* connected to the output of the preamplifier.

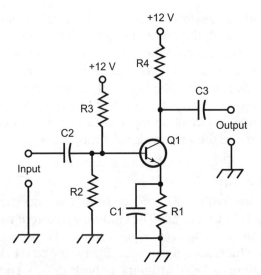

FIG. 5-2. *An audio preamplifier circuit that can be used with the "crystal radio" to produce a signal strong enough to drive a headset (but not a loudspeaker).*

Follow the flow

In the circuit of Fig. 5-2, audio AC passes from the input through C2 to the base of Q1. The capacitor C2 keeps the power supply's DC from affecting the behavior of the previous circuit (the "crystal radio" set itself). The tiny current variations at the base of Q1 cause larger current variations through the transistor. The amplified audio signal passes through C3 to the output. The resistors govern the current flowing through Q1, and must be chosen by experiment for optimum amplification. Capacitor C1 keeps the emitter at audio signal ground while allowing some DC voltage to exist there.

Figure 5-3 is a diagram of a circuit that looks, at first glance, more complicated than Fig. 5-1 or Fig. 5-2. But is it really? If you look at Fig. 5-3 for a minute or two, you'll notice that it's nothing more than the composite of the "crystal radio" and the audio preamplifier that we've just seen. The components are renumbered generally going from left to right, the direction of signal flow through the system. (You never want to duplicate a component designator in any schematic.) In Fig. 5-3, the connection between the original "crystal radio" and the preamplifier corresponds to the short, horizontal connecting line that goes from the dot above C2 to the left-hand side of C3.

Now that you can envision the two building blocks that make up the circuit of Fig. 5-3, the whole diagram looks less complicated, doesn't it? You can follow the signal flow through the system by thinking of the signal going through the "crystal radio" and then through the audio preamplifier. Of course, the process goes so fast that it would seem instantaneous if you could actually see the RF and audio impulses. The whole process, from the signal arriving at the antenna to the audio appearing at the output, takes place in a minuscule fraction of a second. Currents in electrical conductors, in general, travel at roughly 10 percent of the speed of light in free space. That's 30,000 kilometers, or 18,600 miles, per second!

The circuit of Fig. 5-3 has a low audio output power level, although it's a lot louder than the feeble audio signal that comes from the "crystal radio" all by itself, which gets its power only from the signal current in the antenna! Nevertheless, even the amplified audio at the output of the circuit of Fig. 5-3 isn't enough to provide a comfortable listening volume in a loudspeaker. In order to boost the audio power level some more, you'll need an *audio power amplifier.* Figure 5-4 shows a two-transistor circuit that will perform this task. It's called a

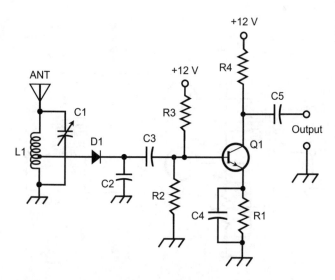

FIG. 5-3. *Combination of "crystal radio" and audio preamplifier circuits. Some of the component designators in the preamplifier stage are updated from Fig. 5-2.*

push-pull circuit. One of the transistors amplifies the "positive" half of the AC audio wave, and the other transistor amplifies the "negative" half. You might say that Q1 does the "pushing" and Q2 does the "pulling," so that when you combine their outputs, you get an amplified version of the complete audio input wave.

The circuit of Fig. 5-4 will accept the output of a low-level audio amplifier and boost it. However, the output from the original "crystal radio" would not be enough to drive the power amplifier. When you hear the term *power amplifier,* you should remember that the circuit does exactly what its name implies. It takes an input signal with a certain amount of *power* and produces an output signal that has *more power.* If the original input signal contains little or no power, then the circuit of Fig. 5-4 won't get enough *drive* (input power) to produce any output signal. In order to work properly, a power amplifier needs a signal that has a modest amount of power to begin with; otherwise it will simply "suck up" the feeble input signal and produce nothing from it other than a tiny amount of heat in the components. The circuit of Fig. 5-3 (the audio preamplifier) provides enough output power to adequately drive a push-pull audio amplifier. The circuit of Fig. 5-1 (the "crystal radio" alone) doesn't.

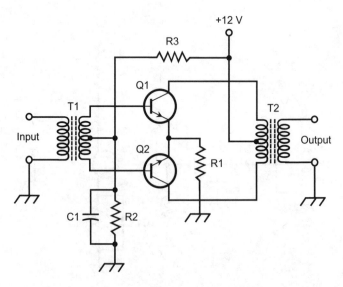

FIG. 5-4. *Audio power amplifier circuit, suitable for driving a speaker or set of speakers.*

Follow the flow

In the circuit of Fig. 5-4, the input signal goes through T1 and appears across its secondary. The details of the ensuing process get rather complicated, involving wave phase and transistor characteristics—technical matters that go beyond the scope of this book. Basically, Q1 handles half of the signal cycle and Q2 handles the other half, as the audio input AC flows alternately up and down through the secondary winding of T1. Transformer T2 combines the two amplified cycle halves back into a complete AC audio wave that's stronger than the wave that came in through T1, thanks to the transistors.

Now if you combine the circuits from Figs. 5-1, 5-2, and 5-4 in *cascade* (one after the other), you get a complete amplitude-modulation (AM) radio receiver that will produce decent sound from a loudspeaker! Figure 5-5 shows the entire three-transistor AM radio receiver in a single schematic diagram. Again, some of the component designators are changed from previous diagrams, so that they increase generally as you go from the original input at the antenna to the final output at the speaker.

So there!

If you were to have seen Fig. 5-5 at the start of this chapter, you might have experienced some major frustration. But now that you can see how the building blocks go together, you know that you don't have to "choke down the whole burger in one gulp." So it's pretty simple after all, isn't it?

The basic process used to make all electronic circuits breaks down into a sequence of combinations. First, the individual components (resistors, capacitors, diodes, and so on) combine to form simple circuits. Then, simple circuits combine to make more complex circuits. After that, complex circuits combine to form complete devices. Several different devices can combine to create a large system. An amateur radio station is a good example of such a system. It might consist of a *transceiver* (transmitter/receiver in a single box), an *antenna tuner*, a computer, an *interface* unit that goes between the computer and the transceiver, and a *speech processor* that goes between the microphone

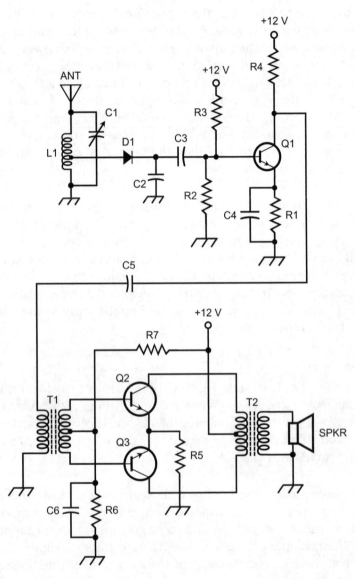

FIG. 5-5. *Complete radio receiver circuit. Some of the component designators in the audio power amplifier stage are updated from Fig. 5-4. The speaker is also shown.*

and the transceiver. Each of these individual devices contains numerous complex circuits, which in turn, comprise multiple simple circuits, which ultimately break down to the individual resistors, capacitors, diodes, transistors, and other components.

Page breaks

Figure 5-5 is a "respectably complicated" diagram. It requires us to draw the system in a "two-story" format with the detector and preamplifier on top, and the audio power amplifier on the bottom. A long, tortuous line, broken in the middle by C5, represents the connection between the preamplifier output and the power amplifier input. There's nothing technically wrong with this diagram, but some people might rather see it all on one level. In order to put Fig. 5-5 all on one level, we'd either have to make it horribly small, or else draw it sideways on the page. But there's another option. We can produce it on a foldout page (the sort of thing that they do in those upscale print magazines when they want to show you something spectacular).

We have yet another alternative, though! We can split the diagram up among multiple pages. That approach isn't necessary here, but when we get to truly complicated systems, such as amateur radio transceivers, television sets, or complete computers, it's an option that engineers often use. Figure 5-6 shows how we can take advantage of this technique with the diagram of our radio receiver. Figure 5-6A puts the detector and the audio preamplifier right-side-up on a single page along with an output designator that appears as an X inside a wedge-like arrow that points off the page toward the right. Figure 5-6B shows the audio power amplifier with an input designator comprising an X inside a wedge-like arrow that points off the page toward the left.

Tip

In Figs. 5-6A and 5-6B, the wedge-like arrows represent points meant to connect directly to each other. They take the place of the long, tortuous line in Fig. 5-5. In this case, we need only one set of arrows of this type. Some circuits need two or more, so we might label them as X, Y, and Z, for example. Then the two X's would connect, the two Y's would connect, and the two Z's would connect.

Let's follow the signal through Fig. 5-6A. A radio wave causes current to flow in the antenna, and also through the inductor L1. Capacitor C1 causes the inductor/capacitor combination (called an *LC circuit,* where L stands for inductance and C stands for capacitance) to resonate at the frequency of the radio signal that we want to hear. Diode D1 detects the RF signal, splitting the AF and the RF portions apart. Capacitor C3 passes the AF part of that signal along to the base of transistor Q1. Capacitor C2 shunts the RF portion of the diode's output to ground because the circuit doesn't need the RF energy anymore, and its presence would only cause trouble. Transistor Q1 acts as an amplifier for the extremely weak AF signal at its base. Resistors R1, R2, R3, and R4 ensure that Q1 gets optimum DC voltage (called *bias*), so that it will produce the greatest possible amount of amplification. Capacitor C4 keeps the emitter at AF signal ground, while allowing some DC voltage to exist there. The AF output signal, along with some DC from the power supply (+12 V), goes off the page through the rightward-pointing arrow marked X.

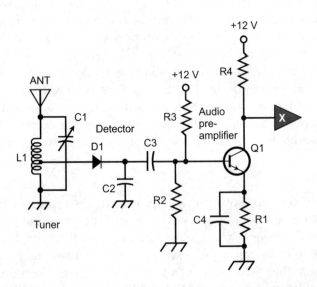

FIG. 5-6A. *Tuner, detector, and audio preamplifier stages in the radio receiver. The wedge X represents an extension to illustration B on the next page.*

Now let's look at Fig. 5-6B and follow the signal after it comes in from the previous diagram. The AF energy, along with some DC, appears at the leftward-pointing arrow marked X. Capacitor C5 blocks the DC so only the AF wave gets to potentiometer R8, which serves as a volume control. The full AF voltage appears across the entire resistance of R8, represented by its left-hand and right-hand terminals. The slider "picks off" various AF voltages that can range from zero (all the way to the right-hand end of the zig-zag, at ground) to the full audio voltage (all the way to the left-hand end of the zig-zag, at C5). This AF voltage goes to the primary of transformer T1. From there, the signal flows in precisely the same way as described in "Follow the flow" for Fig. 5-4. The only difference between this situation and that one lies in the numbering of the component designators. We continue going on upward in numbers from the previous part of the system. We also add a speaker to the output of T2, rather than simply labeling it "Output."

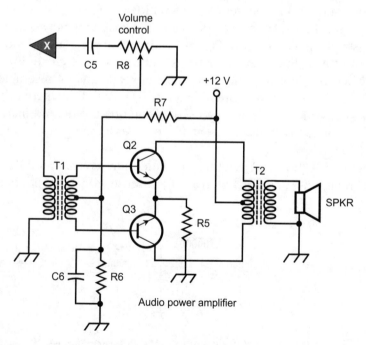

FIG. 5-6B. *Audio power amplifier and speaker in the radio receiver. The wedge X represents an extension from illustration A on the previous page.*

Tip
While looking at Figs. 5-6A and 5-6B, have you noticed that we've added labels to each stage to describe its function? You should do that whenever you draw diagrams of multistage systems such as this one, as long as you have enough room on the page. In Fig. 5-6A we see the labels "Detector" and "Audio preamplifier;" in Fig. 5-6B we see the label "Audio power amplifier."

Some more circuits

Figure 5-7 shows an antenna matching circuit known as an *L network*. Here, the letter L refers to the general layout of the components in the diagram, not the property of inductance. (Actually, to make the coil and capacitor in Fig. 5-7 actually take the shape of an L in the layout, you'll have to rotate the page 90 degrees clockwise and then hold it up to a mirror! But you get the general idea, right?)

Figure 5-8 shows another type of antenna matching network, which consists of the circuit from Fig. 5-7 with an extra capacitor added at the input end. Engineers sometimes call this type of circuit a *pi network* because its components, in the schematic layout, resemble the shape of the upper case Greek letter pi (Π). Both the L network and the pi network are commonly used in antenna tuning and matching circuits for radio transmitters at frequencies ranging up to around 150 MHz.

Figure 5-9 shows a circuit that's a little more complicated than the ones in Figs. 5-7 and 5-8, but in a certain sense, contains both of them

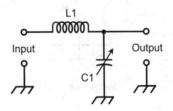

FIG. 5-7. *An L network comprising an inductor and a variable capacitor.*

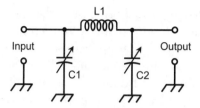

FIG. 5-8. *A pi network comprising an inductor and two variable capacitors.*

put together. When we follow a pi network with an L network, we get a so-called *pi-L network*. The advantage of a circuit like the one in Fig. 5-9, compared to those in the previous two diagrams, lies in its ability to make radio transmitters work with antennas that would otherwise not accept power very well. Those two extra components can go a long way!

Tip

The circuits of Figs. 5-7, 5-8, and 5-9 can all be made more versatile by using variable inductors rather than fixed inductors. One type of variable inductor has become popular among radio amateurs. It's called a *roller inductor*. Just for fun, try an Internet search on "roller inductor" by entering the term in the phrase box of your favorite search engine. You should find some images of these components. They allow for precise adjustment of inductance, and some of them have calibrated crankshafts so that you can easily reset them to any previous position.

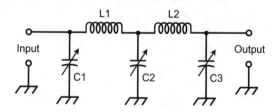

FIG. 5-9. *A pi-L network comprising two inductors and three variable capacitors.*

If you're old enough, you'll remember the days when you had to learn the *International Morse code* (often simply called "the code") to get an amateur radio operator's license. Those days have passed into history, but some radio amateurs still enjoy conversing in this mode. In order to do that, of course, you need to learn the code. To that end, you can build a *code practice oscillator,* such as the one diagrammed in Fig. 5-10. Basically, this circuit is an audio oscillator that you can switch on and off with a Morse code *straight key* or *telegraph key,* labeled "Key" in the figure. (Because a telegraph key technically constitutes an SPST switch, you could label it S1.)

When you first examine Fig. 5-10, you might wonder why an audio oscillator circuit has to be so complicated. Can't you just build a simple audio amplifier, like the preamplifier or power amplifier discussed earlier, and feed some of the output back to the input? Well, yes, you can in fact do that; but if you want a decent audio tone to

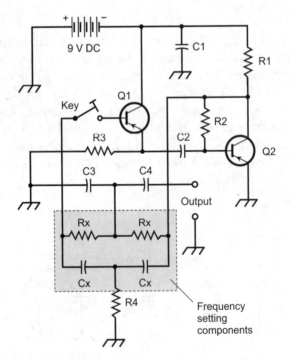

FIG. 5-10. *An audio code-practice oscillator using two PNP bipolar transistors. The values of Rx and Cx determine the frequency.*

come out of your code practice oscillator, you'll get a lot better results with a circuit like the one in Fig. 5-10. This particular circuit is called a *twin-T oscillator* because of the T-shaped configurations including the resistors marked Rx and the capacitors marked Cx. The twin-T oscillator produces an AF tone that's pleasing to the ears, and that also occurs at a predictable and stable *pitch* (frequency).

Follow the flow

The AF signal in the circuit of Fig. 5-10 goes around and around; that's how oscillation happens. We can start pretty much anywhere and follow the signal back to the same point. If we begin at the key, the signal goes into the base of Q1, where it's amplified and undergoes *phase inversion* (the wave is turned upside down). The output of Q1 comes from the emitter of Q1, rather than from the collector; that's a matter of engineering choice because it provides a stable and reliable circuit. The signal then goes into the base of Q2, where it's amplified and inverted again, so it comes out of Q2 in *phase coincidence* (reinforcing rather than opposing) with respect to the signal at the key. The signal emerges from the collector of Q2 and gets routed down to the twin-T network comprising resistors Rx and capacitors Cx, whose values determine the frequency of oscillation. From there, the signal goes back to the key, and gets ready for another round trip through the whole circuit! We take the output through C4, from the point between the two resistors Rx.

The circuit of Fig. 5-10 uses a 9-V battery as its power source. Note that the transistors are of the PNP type, so the collectors get a negative voltage, while the positive battery terminal goes straight to ground. This circuit, therefore, constitutes a *positive-ground system.* We might build a power supply designed to produce -9 V DC from the AC utility mains (about 120 V AC at 60 Hz in the United States), rather than relying on a battery. If we decide to do that, we'll need to make sure that we design the power supply so that it produces a negative voltage, not a positive voltage, with respect to ground.

Figure 5-11 shows a power supply that will do a good job of providing a constant, pure −9V DC. It's just about the same circuit as the one you saw back in Chap. 4 at Fig. 4-13. The basic component

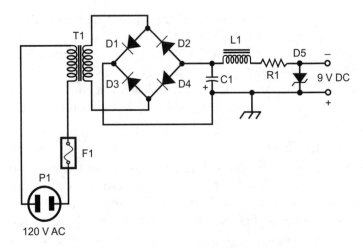

FIG. 5-11. *A regulated −9 V DC power supply that can be used with the code-practice oscillator. Note the positive ground in this supply, designed for use with PNP transistor circuits.*

layout is the same, except that all the diodes go in the opposite direction in Fig. 5-11 as compared to Fig. 4-13 (including the Zener diode), and the electrolytic capacitor polarity is reversed. And of course, the voltage of the power supply shown in Fig. 5-11 is lower than the voltage of the supply shown in Fig. 4-13.

Figure 5-12 shows a complete code practice oscillator system that will operate from the AC utility mains. It includes the power supply from Fig. 5-11 along with the oscillator from Fig. 5-10, all in one big schematic. As we did with the radio receiver circuit diagram of Fig. 5-5, we have connected the power supply output to the oscillator in Fig. 5-12 with a single line, although not as long as the one in Fig. 5-5. We've also included a volume control and a pair of headphones at the output of the twin-T oscillator circuit.

We can apply the same general layout tactics to Fig. 5-12 as we used with Fig. 5-5. We might draw the diagram with the page tilted sideways (landscape orientation instead of portrait orientation), putting the power supply on the left and the oscillator on the right. Or we might use a foldout page. However, an even better alternative (in my opinion) is to split the diagram up into multiple pages so we can draw everything in an uncluttered, right-side-up fashion.

Let's take this business a step further. We can boost the output of the twin-T oscillator so that it will drive a loudspeaker, letting us

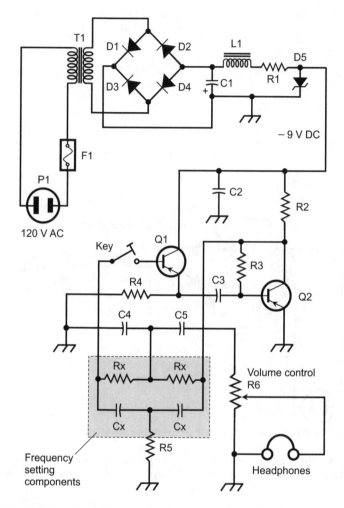

FIG. 5-12. *Combination of the regulated power supply and code-practice oscillator. Note the addition of the volume control and headphones.*

send code to a classroom full of eager Morse-code students! (Finding enough people interested in learning the code to fill up a whole classroom would constitute an entirely different challenge.) A push-pull audio amplifier, just like the one we used for the radio receiver (Fig. 5-4) except with PNP rather than NPN transistors, will work here. Now we have a system with three essential circuits: a power supply, an oscillator, and an amplifier. Figures 5-13A, 5-13B, and 5-13C show how we can draw the complete schematic diagram of this system, letting it spread across three different pages.

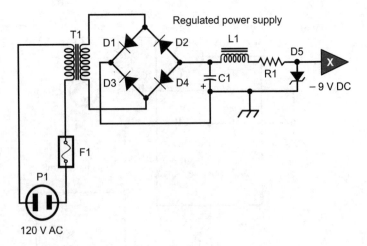

FIG. 5-13A. *Regulated power supply for a classroom code-practice system. The wedge X represents an extension to illustrations B and C on the next two pages.*

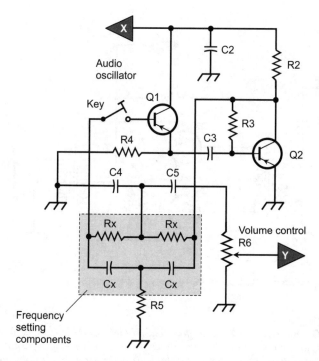

FIG. 5-13B. *Twin-T audio oscillator for a classroom code-practice system. The wedge X represents an extension from illustration A on the previous page. The wedge Y represents an extension to illustration C on the next page.*

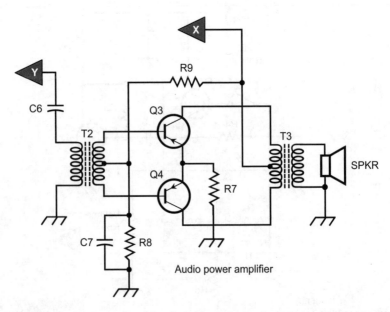

FIG. 5-13C. *Audio power amplifier for a classroom code-practice system. Note the PNP transistors, consistent with the negative power-supply voltage (positive-ground system). The wedge X represents an extension from illustration A on the page before last. The wedge Y represents an extension from illustration B on the previous page.*

Figure 5-14 shows a simple *inductance-capacitance* (LC) **circuit**. Note its resemblance to the L network of Fig. 5-7. In this case, the inductor and capacitor have been reversed from their relative positions in the earlier case. In addition, this circuit has a different purpose than the other one does. The circuit in Fig. 5-7 works mainly for the purpose of tuning an antenna system, or matching the output of a transmitter to the characteristics of a particular antenna. The circuit in Fig. 5-14 is designed to let signals pass through (or not) depending on their frequency. In this case, we have a so-called *highpass filter* because it lets signals go through more easily as the frequency increases. The exact frequency at which the transition from high attenuation (or lots of signal loss) starts to change over to low attenuation (little or no signal loss) depends on the values of the capacitor and inductor.

Figure 5-15 shows a more complicated LC circuit that consists of two filters, one after the other. The first filter, made up of capacitor C1 and inductor L1, is the same highpass filter design as the one in Fig. 5-14. The second LC combination, made up of inductor L2 and capacitor C2, forms a *lowpass filter*. It works in the opposite manner from a highpass filter, letting signals through more easily as the frequency goes down. When we follow a highpass filter with a lowpass filter, and if we choose the *cutoff frequencies* (or transition points)

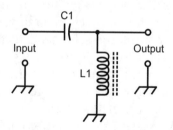

FIG. 5-14. *A simple frequency-sensitive filter circuit.*

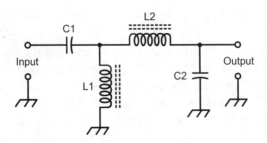

FIG. 5-15. *A complex frequency-sensitive filter comprising two simple, but different, filters connected in cascade.*

so that the highpass filter's cutoff frequency lies below the lowpass filter's cutoff frequency, we can get a *bandpass filter* in which a signal can get through easily only when its frequency lies between the two cutoffs.

> ### Tip
> You'll often find repetition of circuit configurations, one after another, in electronic system design. Sometimes you'll see several identical or similar circuits connected together. Sometimes their components will all have the same values, and sometimes they won't. Sometimes the circuits will be connected in series (end to end, like the links in a chain); sometimes they'll be connected in parallel (across each other, like the rungs in a ladder).

If you know how one circuit in a repetitive system operates, then you know, by extension, how all of the circuits work. A problem that occurs in one circuit might also occur in any of the others, and you can use the schematic diagram to track down the problem. For example, if, through testing, you know that an oscillator has changed frequency because of a defective resistor in the base circuit, then if another oscillator of the same configuration develops the same malfunction, you can consult the schematic, locate the base resistor, and conduct some tests to see if it, too, has gone bad. Without the schematic, you'd have a hard time locating the proper resistor. This single miscreant component that costs a few pennies, could, all by itself, bring down a sophisticated system!

Breaking a massive system down into complex circuits makes it easier to determine which circuit might be creating the problem for the entire system. This complex circuit is then broken down into simple circuits, and a determination is made as to which simple circuit might be at fault. The simple circuit is further divided into separate components, which are examined individually for a possible fault. Through this systematic method of elimination (all done with the help of schematics), the equipment can be repaired with a lot less trouble than you'd experience if you didn't have the schematics. The process of narrowing-down a problem zone from the system to a complex circuit, then to a simple circuit, and finally to an individual component, is called *troubleshooting to the component level.*

You're hired!

If you can troubleshoot large electronic systems to the component level, you'll find yourself in high demand. All you'll have to do is show up on time when someone calls you for service, do the job, and leave. When word gets around about your competence and reliability, you'll have a job for the rest of your life.

Many failures in complex electronic systems arise from a problem with a single component. Sometimes this failure will cause other components to go bad as well, but the repair must start with the first failure. Occasionally, two simultaneously defective components will be discovered, but this sort of coincidence is a rarity. Certainly, you must become familiar with the equipment you intend to repair, but after you have studied and understood the system's normal state of operation, you can use a schematic drawing to identify the general area of any future faults. By following this procedure, you can identify the most suspect individual component(s). Then all you have to do is find the suspect component(s) in the physical system, get into the equipment with the appropriate test instrument, and check the suspect component(s) one by one.

Tip

Even if you have good test equipment, you'll find it difficult to quickly identify defective sections (and especially individual components) without a schematic diagram because you probably won't know where on the circuit board or chassis to look!

Getting comfortable with large schematics

No matter how much you love electronics, you can't expect to sit down as a rank beginner and read complicated schematics straight away. You've got to climb a learning curve. First of all, you must make certain that you know all of the schematic symbols that you expect to deal with. Complex schematics can serve as a great learning tool because they contain lots of symbols, some of which you probably won't know. Although you can't sit down with a complex schematic at the start and understand everything that happens in the circuit that it represents, you can use such diagrams to help you learn the symbols.

Once you feel comfortable with the individual symbols, put away the complex schematics and start looking over diagrams of basic, simple, widely used circuits. You'll find lots of them in magazines geared to electronics enthusiasts. (You'll find some simple projects and related diagrams in *Electricity Experiments You Can Do at Home*, published by McGraw-Hill, as well!) Don't study only one type of schematic, such as those that portray only amplifiers. Check into oscillators, power supplies, solid-state switches, RF circuits, audio circuits, and anything else you can find. You'll discover similarities between different types of circuits, sometimes with no major differences other than a few changes in the component values. When you can identify an amplifier or oscillator or detector circuit by looking at its schematic, then you'll know you're making progress.

Once you can comfortably identify simple electronic circuits from schematic diagrams, you should move on to more complex drawings—but don't go overboard here! If you try to move too quickly, you might grow frustrated and give up altogether. Your next step will include devices that combine a few of the simple circuits you have previously studied. Sometimes additional components are added to match the output of one circuit to the input of another. Select books and publications that offer both theoretical and practical discussions of the circuit that the schematic depicts. Even better, you can build some simple circuits in a workshop at home.

You'll likely get a surprise when you see how your first "homebrewed" electronic circuit looks when compared with the schematic drawing. Your study will continue from this point by examining the functional circuit and noting the relationship of the physical components to those in the schematic drawing. You can further your electronics knowledge by experimenting with these circuits (substituting different components, for example). You might find a way to improve the circuit operation. All improvements should be duly noted, and a new schematic can be drawn up indicating your changes. You might want to simply pencil in the changes on the schematic diagram you were building from.

When you feel comfortable building electronic circuits from simple schematic diagrams, you might want to combine two or more circuits to make a single, more sophisticated device. Take two schematic diagrams from a projects book and combine them on paper. You will have to draw your own schematic to serve as your plan for the building procedure. You might know enough by this time to design and build some additional circuits that can *interface* the two (connect the output of the first circuit to the input of the second one so they both work at their best). When you combine electronics circuit-building with the task of learning to read and create schematics, you can improve your electronics knowledge faster than you can by simply looking at and drawing theoretical diagrams.

Tip
Admittedly, you might start to grow bored as you pore over schematic diagrams for hours on end. But when you can refer to a portion of a schematic drawing and then wire the components in place, much of the boredom will go away. If you're lucky, or if you happen to be a budding *technophile* (technology lover), the stuff will get downright fascinating!

Before you know it, you'll have obtained a solid basic knowledge of schematic diagrams and circuit-building. The circuits that you once imagined as complicated will soon seem familiar and elementary. Nevertheless, you should exercise some caution here. You might feel the temptation to stay with the sorts of circuits that you know best, and not venture into new territory. Don't let that little bit of laziness get the better of you! As soon as you reach one stage of comfort, move on to diagrams that are more difficult and make you feel a little bit uncomfortable again. If you don't "push yourself" this way, you'll find yourself stuck at one stage of development for a long time. Try building more and more complicated devices. Admittedly, this practice can grow expensive if you overdo it; so if you can't build everything in sight, keep on reading schematics and deciphering the various and diverse circuit components anyway.

You'll never cease to be amazed at what you know and what you don't know. For instance, many people new to electronics feel that a commercial AM radio transmitter must constitute a highly complex system. Most electronics novices are astonished to learn that the AM transmitter is technically less complex than an old-fashioned transistorized pocket receiver that you might use to intercept the broadcasts. A commercial radio transmitter is a rather simple system, even though it might have the size and mass of a full-sized refrigerator. The transmitter size is directly related to the component size, which in turn is directly related to the amount of power that the system consumes. The power-supply transformer for a commercial transmitter, all by itself, might weigh as much as the aforementioned refrigerator! This and other components make the commercial broadcast transmitter necessarily large. However, the power-supply transformer for a small amateur radio transmitter will likely weigh less than a kilogram. Schematically, both transformers look the same in schematic diagrams. Circuit complexity bears no relation to the physical size of a component. Complexity depends on the number of components as well as on the number of circuits that the system has.

The massive pieces of equipment are rarely the most complicated ones, both electronically and schematically. The tiny units that can be held in the palm of your hand often take the prize for schematic complexity when you break them down to the component level. A tablet computer is an excellent example. The integrated circuits (ICs or chips) inside a device like that contain millions of individual diodes, transistors, capacitors, and resistors. For this reason, the beginner to electronics and schematics should not shy away from any particular circuit, device, or equipment just because its size suggests complexity. You might be wrong, but even if you're right, every schematic diagram will contain portions that you can comprehend.

After you have passed the intermediate stage of learning schematics, then you can tackle complex circuits, devices, and systems. You can break them down into multiple-circuit stages or devices, and ultimately into simple circuits. Try to obtain schematics of a complex nature that also include a thorough explanation of how the circuits work.

Recall the block diagram of Fig. 2-2, the strobe light circuit that we saw in Chap. 2. Compare it to Fig. 5-16, a schematic that shows all of

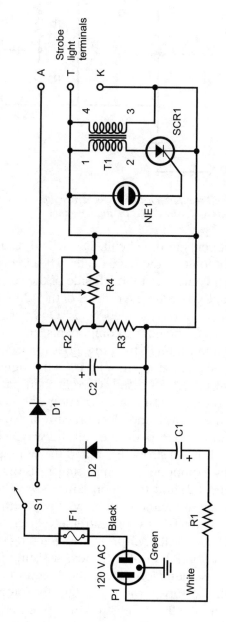

FIG. 5-16. *Complete strobe light circuit originally shown in the block diagram of Fig. 2-2 (Chap. 2). In order to fit it on the page, the whole diagram is rotated 90 degrees.*

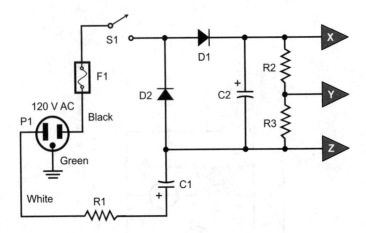

FIG. 5-17A. *Plug, fuse, and rectifier portions of the strobe light circuit. Wedges X, Y, and Z represent extensions to illustration B on the next page.*

the individual components in the circuit. The whole diagram is tilted on its side, allowing it to fit on the page neatly. The circuit is powered with 120 V AC, which comes in at the left side of the schematic (after you've rotated the page). The three terminals of the 120 V AC line take three separate paths along color-coded wires. One wire goes to the fuse, another wire goes to the timing components, and the third wire, coming from the ground prong of the plug, goes to electrical ground.

Following the top signal path, current passes through the fuse F1, assuming that the switch S1 is closed. Current then passes through the rectifier diodes D1 and D2 in one direction only. Conventional current travels with the arrows, and electron current goes against the arrows. From the diodes, one path goes all the way across the top wire to the A terminal of the strobe light, and the other path goes to the adjustable timing components comprising resistors R1 through R4, capacitors C1 and C2, and the neon lamp NE1. The adjustable timing components determine the frequency at which the light operates (how long it stays off between extremely brief flashes). The silicon-controlled rectifier SCR1 performs the switching operation for the strobe lamp. Resistors R2 and R3, besides assisting with the timing operation, provide a junction where the bottom portion of the circuit interacts with the top portion, sending the proper signal to the transformer T1 and finally to the strobe lamp terminals T and K.

Figures 5-17A and B show the same circuit as Fig. 5-16 does, except that the diagram is split between two pages so that it can all

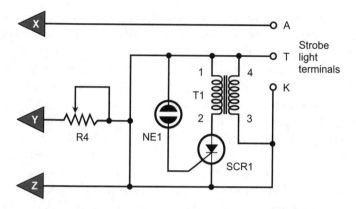

FIG. 5-17B. *Timing and transformer portions of the strobe light circuit. Wedges X, Y, and Z represent extensions from illustration A on the previous page.*

be presented right-side-up instead of sideways. The first part (Fig. 5-17A) shows the power supply and part of the timing circuitry. The second part (Fig. 5-17B) shows the frequency-adjusting potentiometer R4 along with the rest of the timing circuitry, the switching device, and the transformer that provides the strobe light with the voltage it needs to operate properly.

Summary

Reading and drawing schematic diagrams involves breaking down complex circuits into simple ones. The secret is to look at the system's parts and how they relate to each other, rather than to regard the whole thing as a huge monolithic entity. As you systematically study a complex schematic drawing, the relationships between the circuits will get more and more apparent. Once in awhile, you'll suddenly see all of a system's "secrets" revealed at once: An "Aha" moment!

Tip
When you use schematics for electronics troubleshooting, you won't always have to understand the function of every single system element. In many cases, you'll only have to worry about the circuits or components that represent potential trouble spots.

Learning to read and write schematic diagrams is a lot like learning to receive and send the old Morse Code. Morse Code is a language of audible symbols, just like schematic drawing is a language of printed symbols. Once you learn either language, you can use it to communicate; Morse code communicates words and sentences, while schematics communicate principles and concepts.

Using Morse Code as a further example, a long sequence of dots and dashes will mean nothing unless you can break the data down into words. As your proficiency increases, you'll stop hearing mere dots and dashes (or, as some people say, dits and dahs) and hear letters of the alphabet instead. As you keep practicing with the code, you'll start to hear entire words. Eventually, if you keep at it long enough (and especially if you get fond of the code for its own sake, communicating with it for hours on end, as I have done over the years as an amateur radio operator), you'll hear whole phrases and sentences. It will have become a complete language for you.

Reading and writing schematic diagrams follows a similar pattern of development. At first you'll see individual component symbols. Later, you'll begin to see simple circuits hidden within complex circuits. Then you'll be able to identify and decipher complex circuits. Finally, you'll begin to envision entire systems. This knowledge will not come quickly, but your proficiency will improve every time you practice, if you keep pushing yourself (gently, of course) into new knowledge zones.

6

Let's learn by doing

Earlier in this book, I mentioned a volume called *Electricity Experiments You Can Do at Home*. In this chapter, I've adapted a few of the experiments from that book along with layouts and schematics. You should find these activities entertaining and informative in their own right. As you go, you'll "automatically" bolster your proficiency in reading and interpreting schematics. Let's start with some setup details, including a parts list and the construction of a simple circuit-testing board. Then, if you want, you can do the experiments here. If you enjoy them, then you can buy *Electricity Experiments You Can Do at Home* and do a whole lot more!

Tip

You don't have to buy all the parts, build the breadboard, and do the experiments described here, if you aren't in the mood. I think you'll have more fun and learn things better that way, but if you don't want to spend money and time on "nuts and bolts stuff," you can learn a fair amount by simply following along and conducting the experiments in your imagination.

Your breadboard

Every experimenter needs a good workbench. Mine consists of a piece of plywood, weighted down over the keyboard of an old upright piano, and hung from the cellar ceiling by brass-plated chains. Yours doesn't have to be that exotic, and you can put it anywhere as long as it won't shake or collapse. The surface should be made of a non-conducting material such as wood, protected by a plastic mat or a small piece of close-cropped carpet (a doormat works great). A desk lamp, preferably the "high-intensity" type with an adjustable arm, completes the arrangement.

Before you begin any of the tasks described in this chapter, buy a good pair of safety glasses at your local hardware store. Wear the glasses at all times as you play around with the hardware. Get into the habit of wearing safety glasses whenever you work on electrical or electronic circuits, whether you think you need them or not. You never know when a little piece of wire will go flying when you snip it off with diagonal cutters, or a spark will fly right at one of your eyeballs!

> **Tip**
>
> Table 6-1 lists the items you'll need for the experiments in this chapter. You can find many of these components at Radio Shack retail stores, or you can order them through the Radio Shack website. A few of them are available at hardware stores and department stores.

For the experiments described in this chapter, you'll need a prototype-testing circuit board called a *breadboard*. I patronized a local lumber yard to get the wood for my breadboard. I found a length of "12-inch by 3/4-inch" pine in their scrap heap. The actual width of a "12-inch" board is about 10.8 inches or 27.4 centimeters, and the actual thickness is about 0.6 inch or 15 millimeters. They didn't charge me anything for the wood itself, but they demanded a couple of dollars to make a clean cut so I could have a fine rectangular piece of pine measuring 12.5 inches (31.8 centimeters) long.

Using a ruler, divide the breadboard lengthwise at 1-inch (25.4-millimeter) intervals, centered so as to get 11 evenly spaced marks. Do the same going sideways to obtain nine marks at 1-inch (25.4-millimeter) intervals. Using a ball-point or roller-point pen, draw lines parallel to the edges of the board to obtain a grid pattern. Label the grid lines from A to K and 1 to 9, as shown in Fig. 6-1. That'll give you 99 intersection points, each of which you can designate with a letter-number pair, such as D-3 or G-8.

Table 6-1
Components list for simple electricity experiments. You can find these items at retail stores throughout the United States. Abbreviations: AWG = American Wire Gauge, A = amperes, V = volts, W = watts, and PIV = peak inverse volts.

Quantity	Store type or Radio Shack part number	Description
1	Lumber yard	Pine board, approx. 10.8 × 12.5 × 0.6 inches
1	Hardware store	Pair of safety glasses
1	Hardware store	Small hammer
12	Hardware store	Flat-head wood screws, 6 × 3/4
100	Hardware store	Polished steel finishing nails, 1-1/4 inches long
1	Department store	12-inch plastic or wooden ruler
1	Hardware store	Small tube of contact cement
1	Hardware store or Radio Shack	Digital multimeter, GB Instruments GDT-11 or equivalent
1	Hardware store	Diagonal wire cutter/stripper
1	Hardware store	Small needle-nose pliers
1	Hardware store	Roll of AWG No. 24 solid uninsulated (bare) copper wire
1	278-1156	Package of insulated test/jumper leads
4	Hardware store	Alkaline AA cells rated at 1.5 V
1	270-391A	Holder for four size AA cells in series
1	271-1111	Package of five resistors rated at 220 ohms and 1/2 W
1	271-1113	Package of five resistors rated at 330 ohms and 1/2 W
1	271-1115	Package of five resistors rated at 470 ohms and 1/2 W
1	271-1117	Package of five resistors rated at 680 ohms and 1/2 W
1	271-1118	Package of five resistors rated at 1000 (1 k) ohms and 1/2 W
1	271-1120	Package of five resistors rated at 1500 (1.5 k) ohms and 1/2 W
1	271-1122	Package of five resistors rated at 3300 (3.3 k) ohms and 1/2 W
1	276-1104	Package of two rectifier diodes rated at 1 A and 600 PIV
2	272-357	Miniature screw-base lamp holder
1	272-1130	Package of two screw-base miniature lamps rated at 6.3 V
1	272-1133	Package of two screw-base miniature lamps rated at 7.5 V

Once you've marked the grid lines, gather together a bunch of 1.25-inch (31.8-millimeter) polished-steel finishing nails. Place the board on a solid surface that can't suffer any damage from scratching or scraping. A concrete or asphalt driveway will serve this purpose. Pound a nail into each of the grid intersection points shown by the black dots in Fig. 6-1 (53 nails in all). Make certain that the nails are made of polished steel, preferably with "tiny heads." The nails must not have any coating of paint, plastic, or other electrically insulating material. Each nail should go into the board just far enough so that you can't wiggle it around. I pounded every nail down to a depth of approximately 0.3 inch (8 millimeters), a distance amounting to halfway down through the board.

Using 6 × 32 flat-head wood screws, secure the two miniature lamp holders to the board at the locations shown in Fig. 6-1. Using short lengths of thin, solid, bare copper wire, connect the terminals of one lamp holder to breadboard nails A-2 and D-1. Connect the terminals of the other lamp holder to nails D-2 and G-1. Wrap the wire tightly at least twice, but preferably four times, around each nail. Snip off any

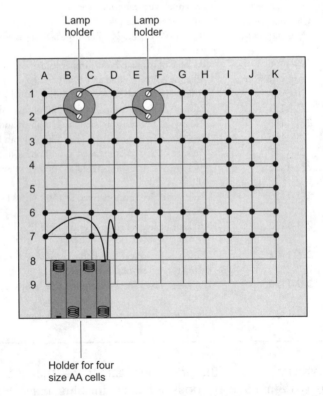

FIG. 6-1. *Layout of the breadboard for simple electricity experiments. I used a "12-inch" pine board (actually 10.8 inches wide) with a thickness of "3/4 inch" (actually about 0.6 inch), cut to a length of 12.5 inches. Solid dots show the positions of the nails. Grid squares measure 1 inch by 1 inch.*

excess wire that remains. Glue the four-cell AA battery holder to the breadboard with contact cement. Allow the cement to harden. That process will need a few hours, so you can take a break for awhile!

When the contact cement has solidified, strip 1 inch of the insulation from the ends of the cell-holder leads and connect the leads to the nails, as shown in Fig. 6-1. Remember that the red lead goes to the positive battery terminal, and the black lead goes to the negative terminal. Use the same wire-wrapping technique that you used for the lamp-holder wires. Place four brand new AA alkaline cells in the holders with the negative terminals against the springs. Now you have a 6-V battery, and the breadboard awaits your exploits.

Wire wrapping

The following breadboard-based experiments employ a construction method called *wire wrapping*. Each nail forms a terminal to which you can attach several component leads or wires. To make a connection, wrap an uninsulated wire or lead around a nail in a tight coil. Make at least two, but preferably four or five, complete wire turns, as shown in Fig. 6-2.

When you wrap the end of a length of wire, cut off the excess wire after wrapping. For small components, such as resistors and diodes,

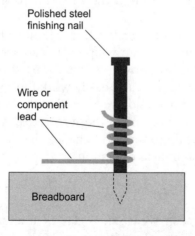

FIG. 6-2. *Wire-wrapping technique. Wind the wire or component lead at least twice, but preferably four or five times, around the nail. Extra wire should be snipped off, if necessary, using a diagonal cutter.*

wrap the leads around the nails as many times as is necessary to use up the entire lead length. That way, you won't have to cut down the component leads. You'll be able to easily unwrap and reuse the components for later experiments. Needle-nose pliers can help you to wrap wires or leads that you can't wrap with your fingers alone.

When you want to make multiple connections to a single nail, you can wrap one wire or lead over the other, but you shouldn't have to do that unless you've run out of nail space. Each nail should protrude approximately 1 inch above the board surface, so you won't be cramped for wrapping space.

Tip

Again, you must make absolutely sure that the nails are made of polished steel *without any coating*. They should be new and clean, so they'll function as efficient electrical terminals.

When you perform the experiments that follow, the exact arrangement of parts on the breadboard is up to you. I've provided schematic and pictorial layout diagrams to show you how the components are interconnected. I recommend that you follow my layout suggestions, only because that way, you can focus on how the actual appearance of the circuit compares with the schematic diagram, even if you haven't bothered to build the breadboard and work with the hardware directly.

Small components such as resistors should always go between adjacent nails, so that you can wrap each lead securely around each nail. *Jumper wires* (also known as *clip leads*) should be secured to the nails so that the "jaws" can't easily work their way loose. It's best to clamp jumpers to nails sideways, so that the wires come off horizontally. If you try to put one of these so-called *alligator clips* down on a nail vertically, there's a good chance that it will pop off in the middle of a mission-critical operation!

Caution!

Please let me repeat: Wear safety glasses at all times as you perform these experiments, whether you think you need to or not.

Kirchhoff's current law

In this experiment, you'll construct a network that demonstrates one of the most important principles in DC electricity. You'll need five resistors: two rated at 330 ohms, one rated at 1000 ohms (1 k), and two rated at 1500 ohms (1.5 k). You'll also need four AA cells.

Mount the resistors on the breadboard by wire-wrapping the leads around the terminal nails, as shown in the layout diagram, Fig. 6-3. Test each resistor with your multimeter (set to work as an ohmmeter) to verify their ohmic values before you install them. Use a 5-inch length of bare copper wire to interconnect the three terminals I-1, J-1, and K-1. Do the same thing with I-3, J-3, and K-3, and also with I-5, J-5, and K-5.

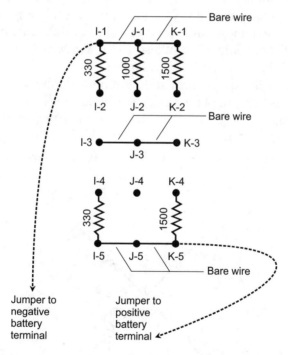

FIG. 6-3. *Arrangement of resistors on breadboard for demonstration of Kirchhoff's current law. All resistance values are in ohms. Solid dots indicate breadboard terminals. Solid lines show interconnections with bare copper wire. Dashed lines indicate jumpers.*

Gustav Robert Kirchhoff (1824–1887) did research and formulated theories in a time when no one knew much about electrical current. He used common sense to deduce fundamental properties of DC circuits. Kirchhoff reasoned that the current going into any *branch point* in a circuit must always equal the current going out of that point. Figure 6-4 shows a generic example of this principle, known as *Kirchhoff's first law*. We can also call it *Kirchhoff's current law* or the *principle of conservation of current*.

Mathematically, the sum of the currents entering a branch point always equals the sum of the currents leaving that same branch point. In the example of Fig. 6-4, two branches enter the point and three branches leave it, so

$$I_1 + I_2 = I_3 + I_4 + I_5$$

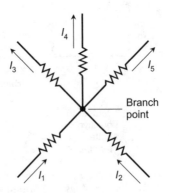

FIG. 6-4. *According to Kirchhoff's current law, the sum of the currents flowing into any branch point is equal to the sum of the currents flowing out of that branch point. In this example,* $I_1 + I_2 = I_3 + I_4 + I_5$.

Kirchhoff's current law holds true no matter how many branches come into or go out of a particular point.

Connect your four-cell battery to the resistive network and measure the currents in each branch. Every test point should be metered individually, while all the other test points are shorted with jumpers. Figure 6-5 is a schematic diagram that shows the actual values of the resistors in my network (yours will be slightly different, of course), along with the value I got when I measured I_1, the current through the smaller of the two input resistors.

As you measure each of the four other current values in turn, make certain that the meter polarity always agrees with the battery polarity. The black meter probe should go to the more negative point, and the red meter probe should go to the more positive point. That way, you'll avoid getting negative current readings that might throw off your calculations. When I tested my four-cell battery to determine its

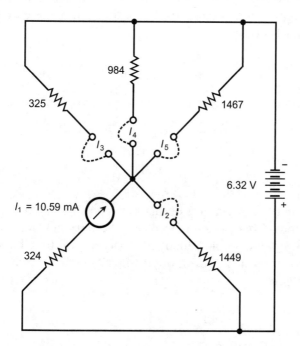

FIG. 6-5. *Network for verifying Kirchhoff's current law. All resistance values are in ohms. The battery voltage, the current* I_1, *and the resistances are the values I measured. Dashed lines show interconnections with jumpers.*

voltage, I got 6.32 V. When I measured I_1 through I_5, I got the following results, accurate to the nearest hundredth of a milliampere (mA):

$$I_1 = 10.59 \text{ mA}$$

$$I_2 = 2.40 \text{ mA}$$

$$I_3 = 8.35 \text{ mA}$$

$$I_4 = 2.79 \text{ mA}$$

$$I_5 = 1.88 \text{ mA}$$

It's "mission critical" that all test points *not* undergoing current measurement be shorted out with jumpers. Otherwise, your network will be incomplete and your current measurements will come out wrong. After you've finished making measurements, remove all the jumpers to conserve battery energy.

Now you can input your numbers to the Kirchhoff formulas and see how close the sum of the input currents comes to the sum of the output currents. Here are my results for the sum of the currents entering the branch point:

$$I_1 + I_2 = 10.59 + 2.40$$
$$= 12.99 \text{ mA}$$

When I added the currents leaving the branch point, I got

$$I_3 + I_4 + I_5 = 8.35 + 2.79 + 1.88$$
$$= 13.02 \text{ mA}$$

Tip

When you do experiments of this sort, you should expect a slight discrepancy. That principle explains the 0.03 mA current difference in the branch points in my test. An error of three hundredths of a milliampere at 13 mA amounts to well under 1 percent, which is acceptable.

Now compare!

Compare the layout diagram (Fig. 6-3) with the schematic diagram (Fig. 6-5) without the meter or jumpers.

Kirchhoff's voltage law

In this experiment, you'll construct a network that demonstrates another important DC circuit rule. You'll need four resistors: one rated at 220 ohms, one rated at 330 ohms, one rated at 470 ohms, and one rated at 680 ohms. You'll also need four AA cells.

According to *Kirchhoff's second law*, the sum of the voltages across the individual components in a series DC circuit, taking polarity into account, always equals zero. We can also call this rule *Kirchhoff's voltage law* or the principle of *conservation of voltage*.

Consider the generic series DC circuit shown in Fig. 6-6. According to Kirchhoff's voltage law, the battery voltage E must equal the sum of the potential differences (voltages) across the resistors, although the polarity will be reversed. Mathematically, we can state this fact as

$$E + E_1 + E_2 + E_3 + E_4 = 0$$

If we measure the voltages across the individual resistors and the battery, one at a time, with a DC voltmeter and *disregard the polarity*, we should find that

$$E = E_1 + E_2 + E_3 + E_4$$

Check each of the four resistors with your ohmmeter to verify their actual values. Mount the resistors in the upper right-hand corner of your breadboard by wire-wrapping the leads around nails, as shown in Fig. 6-7. Connect the battery to the network as shown, and measure the voltage across each resistor. Figure 6-8 illustrates the actual values of the resistors in my network (yours will be slightly different),

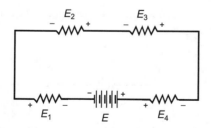

FIG. 6-6. *According to Kirchhoff's voltage law, the sum of the voltages across the resistances in a series DC circuit is equal and opposite to the battery voltage. If we disregard polarity, then in this example, we'll observe that* $E = E_1 + E_2 + E_3 + E_4$.

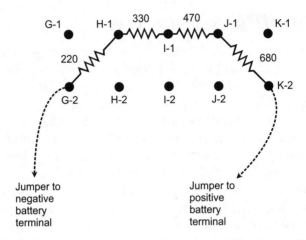

FIG. 6-7. *Suggested arrangement of resistors on breadboard for demonstration of Kirchhoff's voltage law. All resistance values are in ohms. Solid dots indicate terminals. Dashed lines indicate jumpers.*

along with the voltage I got for E_2. I measured $E = 6.30$ V across the battery when it was connected to the resistors.

As you measure each voltage E_1 through E_4, the black meter probe should go to the more negative voltage point, and the red probe should go to the more positive point. That way, you'll avoid getting negative

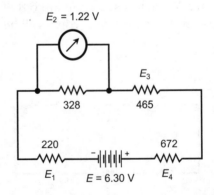

FIG. 6-8. *Network for verifying Kirchhoff's voltage law. All resistance values are in ohms. The battery voltage E, the voltage across the second resistor, and the resistances are the values I measured.*

readings that might throw off your calculations. When I measured the voltages across the individual resistors, I got

$$E_1 = 0.82 \text{ V}$$

$$E_2 = 1.22 \text{ V}$$

$$E_3 = 1.75 \text{ V}$$

$$E_4 = 2.52 \text{ V}$$

When you've finished making your measurements, remove one of the jumpers to take the load off the battery.

After you've double-checked and written down your voltage measurements, input the numbers to the modified Kirchhoff formula

$$E = E_1 + E_2 + E_3 + E_4$$

and see how closely it works out. For the left-hand side of this equation, I measured

$$E = 6.30 \text{ V}$$

In the right-hand side of the foregoing equation, I added my numbers to get

$$E_1 + E_2 + E_3 + E_4 = 0.82 + 1.22 + 1.75 + 2.52$$
$$= 6.31 \text{ V}$$

That's an error of only 0.01 V out of a net potential difference of 6.30 V, amounting to less than two-tenths of one percent error.

Now compare!
Compare the layout diagram (Fig. 6-7) with the schematic diagram (Fig. 6-8), and "follow the flow" through both versions.

A resistive voltage divider

You can use the components from the previous experiment to obtain several different voltages from a single battery. Keep the resistors on the breadboard in the same arrangement as you had them in the experiment for Kirchhoff's voltage law.

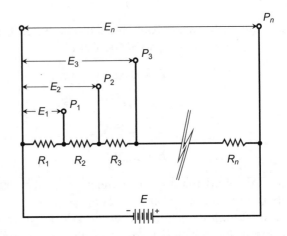

FIG. 6-9. *A voltage divider takes advantage of the potential differences across individual resistors connected in series with a DC power source. Note the use of italics and subscripts for the resistor designators!*

When you connect two or more resistors in series with a DC power source, those resistors produce specific voltage ratios. You can tailor these ratios using specific resistances that "fix" the intermediate voltages. This type of circuit works best when the resistance values are fairly small. Figure 6-9 illustrates the principle of a *resistive voltage divider.* The individual resistances are R_1, R_2, R_3, ..., and R_n. The total resistance R is the sum

$$R = R_1 + R_2 + R_3 + ... + R_n$$

If we call the power supply voltage E, then Ohm's law tells us that the current I at any point in the circuit must be

$$I = E/R$$

as long as we express I in amperes, E in volts, and R in ohms. At the points P_1, P_2, P_3, ..., and P_n, the voltages relative to the negative battery terminal are E_1, E_2, E_3, ..., and E_n, respectively. The last (and highest) voltage E_n is the same as the battery voltage E. The voltages at the various points increase according to the sum total of the resistances up to each point, in proportion to the total resistance, multiplied by the supply voltage. In theory, then, we should find that the following equations hold true:

$$E_1 = ER_1/R$$
$$E_2 = E(R_1 + R_2)/R$$
$$E_3 = E(R_1 + R_2 + R_3)/R$$
$$\downarrow$$
$$E_n = E(R_1 + R_2 + R_3 + \dots + R_n)/R$$
$$= ER/R$$
$$= E$$

Tip

Note the use of italics and subscripts for the resistor designators here! Instead of R1, R2, R3, and so on, we now write R_1, R_2, R_3, and so on. This alternative notation hasn't appeared previously in this book, but lots of engineers use it in schematics. Don't get surprised when you encounter it. This notation is commonly used for capacitors, inductors, diodes, and other components as well.

During this experiment, I measured $E = 6.30$ V across the battery as it worked under load, as shown in Fig. 6-10A. This diagram shows the *rated* values of the resistors. Your actual values will differ slightly from the rated values. In my case, the actual values were

$$R_1 = 220 \text{ ohms}$$

$$R_2 = 328 \text{ ohms}$$

$$R_3 = 465 \text{ ohms}$$

$$R_4 = 672 \text{ ohms}$$

Set your meter to measure current in milliamperes (mA). Connect the battery to the resistive network through the meter, as shown in Fig. 6-10B, and measure the current. In theory, I expected the milliammeter to indicate a value equal to the battery voltage divided by the sum of the actual resistances, or

$$I = E/R$$
$$= 6.30/(220 + 328 + 465 + 672)$$
$$= 6.30/1685$$
$$= 0.00374 \text{ A}$$
$$= 3.74 \text{ mA}$$

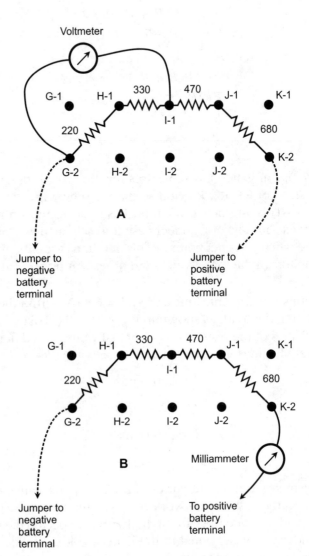

FIG. 6-10. *At A, arrangement for measuring voltages in a resistive divider. Here, the voltmeter is connected to measure the voltage* E_2 *across the first and second resistors. All resistance values are in ohms. Solid dots indicate terminals. Dashed lines indicate jumpers. At B, arrangement for measurement of current through the network.*

When I measured the current, I got 3.73 mA, a value comfortably within the limits of acceptable error.

Now measure the intermediate voltages E_1 through E_4 with your meter set for a moderate DC voltage range. The black meter probe should go directly to the negative battery terminal and stay there. The red meter probe should go to each positive voltage point in turn. First measure the voltage E_1 that appears across R_1 only. Then measure, in order, the following voltages, as illustrated in the schematic of Fig. 6-11:

- The potential difference E_2 across $R_1 + R_2$
- The potential difference E_3 across $R_1 + R_2 + R_3$
- The potential difference E_4 across $R_1 + R_2 + R_3 + R_4$

Figure 6-11 illustrates the arrangement for measuring E_2 to provide a specific example. When I measured each voltage in turn, my meter displayed these results:

$$E_1 = 0.82 \text{ V}$$

$$E_2 = 2.04 \text{ V}$$

$$E_3 = 3.79 \text{ V}$$

$$E_4 = 6.30 \text{ V}$$

After you've finished measuring all the voltages, remove one of the jumpers to conserve battery energy.

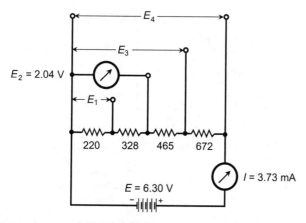

FIG. 6-11. *Network for testing the operation of a resistive voltage divider. All resistance values are in ohms. The battery voltage E, the voltage E_2 across the first and second resistors, and the resistance values represent my measurements.*

Now connect the meter across the combination $R_1 + R_2$. Run a couple of jumper wires to a *load resistor* located elsewhere on the breadboard, as shown in the layout diagram of Fig. 6-12. This arrangement will make the voltage source E_2 force current through the load resistor, which we'll call R_L. Try every resistor in your repertoire in the place of R_L. If you obtained all the resistors in the parts list, you'll have seven tests to do, using resistors rated at values ranging from 220 to 3300 ohms.

Alternately connect and disconnect one of the jumper wires between the voltage divider and R_L, so that you can observe the effect of the load on E_2. As you can see, the external load affects the behavior of the voltage divider. As R_L decreases, so does E_2. The effect becomes

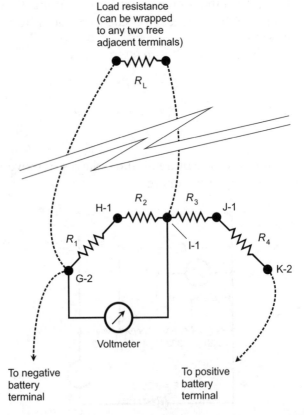

FIG. 6-12. *Circuit for testing a resistive voltage divider under load. Dashed lines indicate jumpers. This diagram shows the arrangement for measuring variations in* E_2 *as the load resistance* R_L *is alternately connected and disconnected from the series combination of* R_1 *and* R_2.

Table 6-2
Here are the voltages that I measured across various loads in a resistive voltage divider constructed according to Fig. 6-12. My network resistor values were R_1 = 220 ohms, R_2 = 328 ohms, R_3 = 465 ohms, and R_4 = 672 ohms. The load resistance values (left column) are the actual measured values for components rated at 3300, 1500, 1000, 680, 470, 330, and 220 ohms, respectively as you read down.

Load resistance (ohms)	Output voltage (volts)
3250	1.83
1470	1.63
983	1.49
671	1.32
466	1.14
326	0.96
220	0.76

dramatic when R_L becomes small, representing a "heavy load." Table 6-2 shows the results I got. Plot your results as points on a coordinate grid with R_L on the horizontal axis and E_2 on the vertical axis, and then "connect the dots" to get a *characteristic curve* showing the voltage as a function of the load resistance. Figure 6-13 is the graph I made.

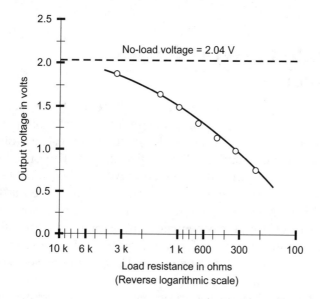

FIG. 6-13. *My results for output voltage versus load resistance in the voltage divider. The dashed line shows the open-circuit (no-load) voltage across the combination of R_1 and R_2 in series. Open circles show the measured voltages under various loads. The solid curve reveals how the circuit behaves as the load resistance goes down.*

I used a *reverse logarithmic scale* to portray R_L, so that the values would be reasonably spread out. This graph scale allows for a graph that provides a clear picture of what happens as the *conductance* of the load *increases.*

Here's a thought!

What do you think will happen to the voltage across the load if you use two, three, four, or five 220-ohm resistors in parallel, getting R_L values of about 110, 73, 55, and 44 ohms respectively? Try the arrangements one by one, and see what happens! Your original package from Radio Shack has five 220-ohm resistors in it, right?

Tip

The results of this experiment suggest that when engineers build voltage dividers, they had better know what sort of external load the circuit will have to deal with. If the load resistance fluctuates greatly, especially if it sometimes gets low, a resistive voltage divider won't work very well.

A diode-based voltage reducer

Rectifier diodes can reduce the output voltage of a low-voltage DC battery or power supply, providing a better (or at least more predictable) way to obtain specific desired voltages than the resistive divider can do. For this experiment, you'll need two diodes. The ones I obtained were rated at 1 A and 600 *peak inverse volts* (PIV), available at Radio Shack stores as part number 276-1104. You'll also need at least one of each of the resistors listed in Table 6-1, along with some jumpers.

Figure 6-14 shows the schematic symbol for a simple rectifier diode, which is manufactured by joining a piece of *P-type* semiconductor material to a piece of *N-type* material. The N-type semiconductor, represented by the short, straight line, forms the diode's *cathode.* The P-type semiconductor, represented by the arrow, composes the *anode.* Under most conditions, electrons can travel easily from the cathode to the anode (in the direction opposite the arrow), but

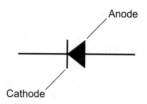

FIG. 6-14. *Schematic symbol for a semiconductor diode. The short line represents the cathode. The arrow represents the anode.*

not from the anode to the cathode (in the direction of the arrow). Conventional current, which always goes from positive to negative, moves in the same direction as the arrow points.

If you connect a battery and a resistor in series with a diode, current will flow if the negative terminal of the battery faces the cathode and the positive terminal faces the anode, as shown in Fig. 6-15A. This condition is called *forward bias.* No current will flow if the bat-

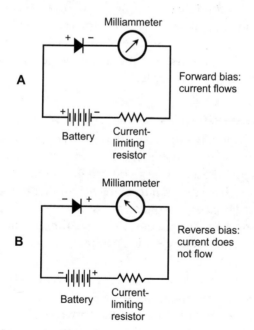

FIG. 6-15. *Series connection of a battery, a resistor, a current meter, and a diode. At A, forward bias causes current to flow if the voltage equals or exceeds the forward breakover threshold. At B, reverse bias results in no current through the diode, unless the voltage gets very high.*

tery is reversed, as shown in Fig. 6-15B (except when the voltage is very high). This condition is called *reverse bias*. The resistor prevents destruction of the diode by excessive current under conditions of forward bias.

It takes a certain minimum voltage to cause current to flow through a forward-biased semiconductor diode. Engineers call this "threshold" the *forward breaker voltage*. In most diodes, it's a fraction of a volt, but it varies somewhat depending on how much current the diode is forced to carry. If the forward-bias voltage across the diode's *P-N junction* is not at least as great as the forward breaker voltage, then the diode will not conduct. When a diode is forward-biased and connected in series with a battery, the voltage goes down to an extent approximately equal to the forward breakover voltage. Unlike the voltage reduction that takes place with resistors, the diode's voltage-dropping capability does not change much when the external load resistance goes up or down.

Although current won't normally flow through a diode that's reverse-biased, exceptions do occur. If the reverse voltage gets high enough (usually far greater than the forward breakover value), a diode will conduct current because of the so-called *avalanche effect*. Zener diodes, which are used to regulate voltages in power supplies, work according to this principle.

Did you know?

When you connect two or more identical rectifier diodes in series with their polarities in agreement, the forward-breakover voltages add up (give or take a little), so you can get stable and predictable voltage drops that come quite close to whole-number multiples of the forward breakover voltage of a single diode. This technique works as long as you connect a load in the circuit, so that the diodes are forced to carry some current.

You can set up a voltage reducer with two diodes in series and their polarities in agreement, as shown in Fig. 6-16, so that current flows through the load resistor R_L if the forward bias is large enough. Figure 6-17 is a pictorial layout diagram showing an arrangement for mounting the components on your breadboard. Set your meter to indicate DC voltage in a moderate range, such as 0 to 20 V. Connect the meter across the load resistance, paying attention to the polarity so you get positive voltage readings. As you did in the previous

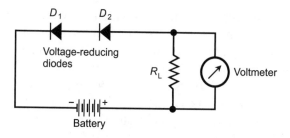

FIG. 6-16. *Schematic diagram showing the method of voltage measurement across a load resistance* R$_L$ *in a two-diode voltage reducer.*

experiment, try every resistor you have for R_L. Measure the voltage across R_L in each case. You'll have seven tests to do, with resistances ranging from 220 to 3300 ohms.

The load resistance R_L affects the behavior of a diode-based voltage reducer, but in a different way than it affects the behavior of a resistive voltage divider. When you do these tests, you'll see that as the load resistance R_L decreases, the potential difference across it goes down, but only a little bit. The voltage across the load tends to drop *more and more slowly* as R_L decreases. Contrast this behavior

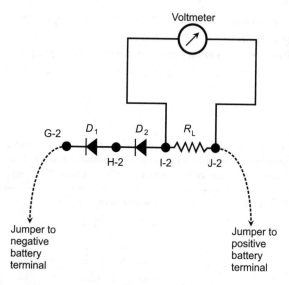

FIG. 6-17. *Suggested breadboard layout for measurements of the voltages across the load resistance in a two-diode voltage reducer. Solid dots show breadboard terminals. Dashed lines indicate jumpers. Pay attention to the diode polarity! The cathodes should go toward the negative battery terminal.*

with that of the resistive divider, in which the voltage drops off *more and more rapidly* as the load resistance goes down. Table 6-3 shows the results I got when I measured the voltages across various load resistances with this two-diode arrangement.

Plot your results as points on a coordinate grid with the load resistance on the horizontal axis and the voltage across the load resistor on the vertical axis, and then approximate the curve as you did in the previous experiment. When I did this little exercise, I got the graph shown in Fig. 6-18. As before, I used a reverse logarithmic scale to portray R_L. Compare this graph with Fig. 6-13 from the previous experiment.

Here's a thought!

Repeat this experiment with only one diode. Then, if you're willing to make another trip to Radio Shack, get another package of diodes and try the experiment with three or four of them in series. You might also obtain some more resistors, covering a range of values of, say, 100 ohms to 100,000 ohms, and test the circuit using those resistors in the place of R_L.

Caution!

Don't use a resistor of less than about 75 ohms as the load here. In this arrangement, a 1/2-watt resistor of less than 75 ohms will let too much current flow, risking destruction of the resistor (and possibly the diodes as well, if you make the resistance really low).

Table 6-3

Here are the output voltages that I obtained with various loads connected to a diode-based voltage reducer. The circuit consisted of two diodes rated at 1 A and 600 PIV, forward-biased and placed in series with a 6.30-V battery.

Load resistance (ohms)	Output voltage (volts)
3250	5.08
1470	4.99
983	4.96
672	4.91
466	4.88
326	4.84
220	4.79

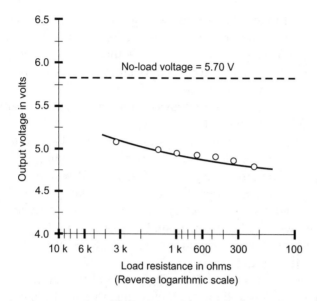

FIG. 6-18. *Output voltage versus load resistance for the diode-based voltage reducer. These are my results. The dashed line shows the open-circuit (no-load) voltage. Open circles show measured voltages under various loads. The solid curve approximates the circuit's characteristic function.*

Mismatched lamps in series

When two dissimilar incandescent lamps operate in series, they receive different voltages and consume different amounts of *volt-ampere* (VA) power, as this experiment demonstrates. (Remember from your basic electricity courses that in a DC circuit, power in watts equals voltage in volts times current in amperes, hence the term *volt-ampere* for simple DC power.) You'll need a 6.3-V lamp, a 7.5-V lamp, four AA cells, and some jumpers.

Your breadboard has two screw-base lamp holders. Position the board so that one holder lies on the left, and the other one lies to its right. Install a 6.3-V lamp in the left-hand socket, and a 7.5-V lamp in the right-hand socket. Connect a short length of bare wire securely between terminals D1 and D2, so that the top terminal of the left-hand lamp holder goes to the bottom terminal of the right-hand lamp holder. Then connect jumpers between the free lamp socket terminals and the battery terminals, so that the lamps are connected in series with each other. Both lamps should glow at partial brilliance.

Call the lamp that's closer to the negative battery terminal "lamp N." That will be the one on the left. Call the lamp that's closer to the positive battery terminal "lamp P." That'll be the one on the right. Unscrew lamp N. At the instant the contact fails, lamp P will go dark. Screw lamp N back in, and then unscrew lamp P. Lamp N will go out. This phenomenon is typical of the behavior of a series circuit. If any component opens up, all the others lose power. A break at any point in a series circuit prevents current from flowing anywhere in that circuit.

Now short out lamp N with a jumper. Lamp N will go dark because it no longer has any voltage across it; lamp P will attain nearly full brilliance. Then disconnect the jumper from lamp N, and move the jumper so that it shorts out lamp P instead. Lamp P will go dark while lamp N glows at full brilliance. Again, this behavior is typical of series circuits. If any component shorts out, all of the others receive more power than they did before.

Set your meter to read DC volts. Connect the jumpers so that the lamps are in series and are both glowing at partial brilliance. Measure the voltage E_1 across lamp N, as shown in the schematic of Fig. 6-19 and in the equivalent layout pictorial of Fig. 6-20. Then measure the voltage E_2 across lamp P, as shown in the schematic of Fig. 6-21 and the equivalent layout pictorial of Fig. 6-22. When I performed these tests, I got the following voltages:

$$E_1 = 2.20 \text{ V}$$
$$E_2 = 3.64 \text{ V}$$

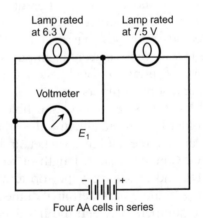

FIG. 6-19. *Measurement of voltage E_1 across the more negative of two dissimilar lamps in series (called N, rated at 6.3 V).*

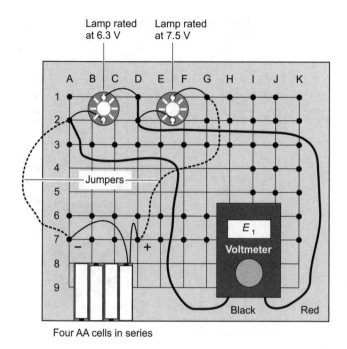

FIG. 6-20. *Breadboard layout rendition of the schematic shown in Fig. 6-19.*

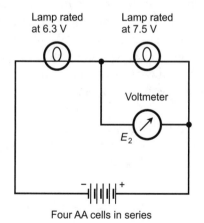

FIG. 6-21. *Measurement of voltage E_2 across the more positive lamp (called P, rated at 7.5 V).*

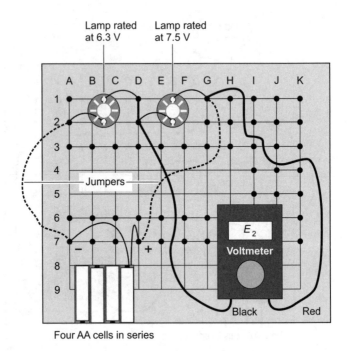

FIG. 6-22. *Breadboard layout rendition of the schematic shown in Fig. 6-21.*

Tip

If you can't get the exact same Radio Shack lamps as specified in the parts list for these experiments (Table 6-1), you can use lamps from another source, such as a hardware store or hobby store. This experiment will work as long as the lamps are rated for slightly different voltages or slightly different power levels, and both are rated between 6 V and 12 V.

Now determine the voltage E across the series combination of lamps, as shown in the schematic at Fig. 6-23 and the layout in Fig. 6-24. In theory, the reading should be

$$E = E_1 + E_2$$

When I put my numerical results into this formula, I predicted that I would see

$$E = 2.20 + 3.64$$
$$= 5.84 \text{ V}$$

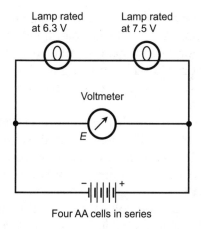

FIG. 6-23. *Measurement of voltage* E *across the combination of two dissimilar lamps in series.*

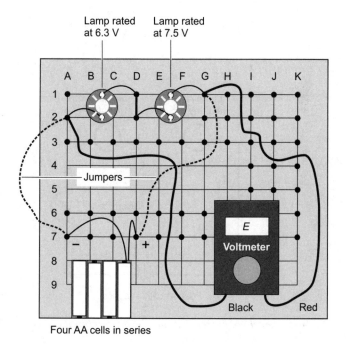

Four AA cells in series

FIG. 6-24. *Breadboard layout rendition of the schematic shown in Fig. 6-23.*

My meter reading was $E = 5.85$ V, a value significantly lower than the 6.30 V I started out with under no-load conditions. Evidently, these bulbs "tax" the four-AA-cell battery rather heavily. I also entertained the notion that because I'd been using the battery for a lot of other experiments besides the ones described here, my AA cells might have grown a little weak.

Set your meter for DC milliamperes, and connect it, as shown in the schematic diagram of Fig. 6-25 and in the equivalent layout pictorial of Fig. 6-26. That connection will allow you to measure the current drawn by the series combination of lamps. I obtained a reading of 139 mA (or 0.139 A).

Now that you know the voltage across each lamp and the current going through the whole circuit, you can determine the VA power figures for the lamps individually and together. Let P_{VA1} represent the VA power consumed by lamp N, and use the formula

$$P_{VA1} = E_1 I$$

When I plugged in my results, I got

$$P_{VA1} = 2.20 \times 0.139$$
$$= 0.306 \text{ VA}$$

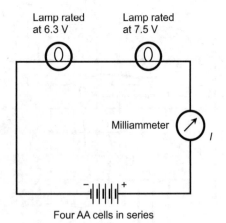

Four AA cells in series

FIG. 6-25. *Measurement of current* I *drawn by the combination of two dissimilar lamps in series.*

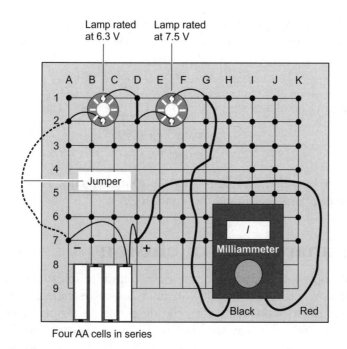

Four AA cells in series

FIG. 6-26. *Breadboard layout rendition of the schematic shown in Fig. 6-25.*

Letting P_{VA2} represent the VA power consumed by lamp P, the formula is

$$P_{VA2} = E_2\, I$$

My result turned out as

$$P_{VA2} = 3.64 \times 0.139$$
$$= 0.506 \text{ VA}$$

Now let's say that P_{VA} represents the VA power consumed by both lamps operating together. In that case, theory predicts that

$$P_{VA} = E\, I$$

Plugging in my experimental results gave me

$$P_{VA} = 5.85 \times 0.139$$
$$= 0.813 \text{ VA}$$

In theory, the VA power consumed by the lamp combination should also work out as the sum of the two VA power quantities taken individually:

$$P_{VA} = P_{VA1} + P_{VA2}$$

Adding my results of $P_{VA1} = 0.306$ and $P_{VA2} = 0.506$, I got

$$P_{VA} = 0.306 + 0.506$$
$$= 0.812 \text{ VA}$$

The error between my two results was a small fraction of one percent, a state of affairs that made me joyful indeed.

Summary and conclusion

When you want to design, build, debug, and troubleshoot electronic equipment, you'll do best if you have a good schematic (or set of schematics) to work with. Pictorial diagrams, including layouts, can help you find your way too. But when it comes down to doing the work, you'll never find any substitute for playing around with real-world hardware. If you have both schematic and pictorial diagrams available, your job will be as easy as it can get.

If you did all the experiments in this chapter, you probably came up with results that differ slightly from mine. If you had to make major parts substitutions, for example with the lamps in the last experiment, then some of your results doubtless came out a lot different than mine did. Whether you did the experiments or merely followed along as a spectator, you got a chance to see how schematics, layout diagrams, literal pictorials, graphs, and tables can work together! All of these tools belong in an engineer's knowledge base.

Again, I'd like to make a "plug" for *Electricity Experiments You Can Do at Home*. You'll get some good hands-on lab experience, combined with some theory and some rather strange phenomena, from that book. If you want a more theoretical and exhaustive presentation of electricity and electronics, with plenty of schematics and just enough mathematics to keep a true geek from getting bored, I recommend the latest edition of *Teach Yourself Electricity and Electronics*. Both books are published by McGraw-Hill, and you can find them at all major book retailers.

A

Schematic symbols

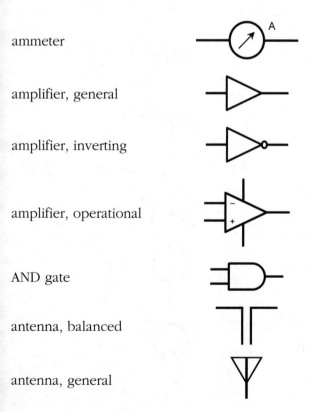

ammeter

amplifier, general

amplifier, inverting

amplifier, operational

AND gate

antenna, balanced

antenna, general

antenna, loop

antenna, loop, multiturn

battery, electrochemical

capacitor, feedthrough

capacitor, fixed

capacitor, variable

capacitor, variable,
    split-rotor

capacitor, variable,
    split-stator

cathode, electron-tube,
    cold

cathode, electron-tube,
    directly heated

cathode, electron-tube, indirectly heated	
cavity resonator	
cell, electrochemical	
circuit breaker	
coaxial cable	
crystal, piezoelectric	
delay line	
diac	
diode, field-effect	
diode, general	
diode, Gunn	
diode, light-emitting	
diode, photosensitive	

diode, PIN

diode, Schottky

diode, tunnel

diode, varactor

diode, Zener

directional coupler

directional wattmeter

exclusive-OR gate

female contact, general

Ferrite bead

filament, electron-tube

fuse

galvanometer

grid, electron-tube

ground, chassis

ground, earth

handset

headset, double

headset, single

headset, stereo

inductor, air core

inductor, air core, bifilar

inductor, air core, tapped

inductor, air core,
   variable

inductor, iron core

inductor, iron core, bifilar

inductor, iron core,
   tapped

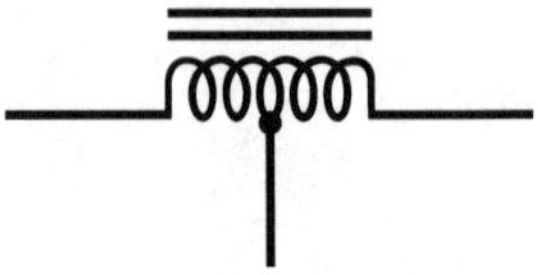

inductor, iron core,
   variable

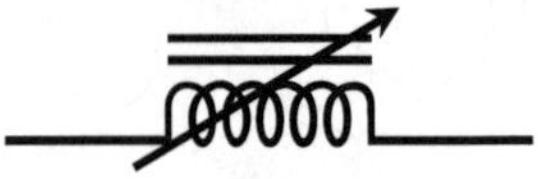

inductor, powdered-iron
   core

inductor, powdered-iron
   core, bifilar

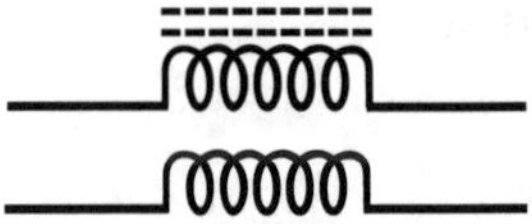

inductor, powdered-iron
   core, tapped

inductor, powdered-iron
   core, variable

integrated circuit, general

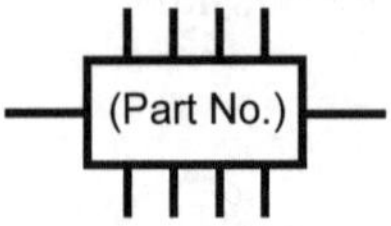

(Part No.)

jack, coaxial or phono

jack, phone, 2-conductor

jack, phone, 3-conductor

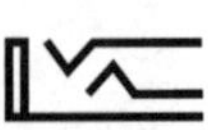

key, telegraph

lamp, incandescent

lamp, neon

male contact, general

meter, general

microammeter

microphone

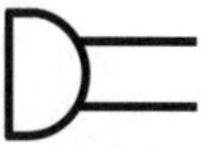

microphone, directional

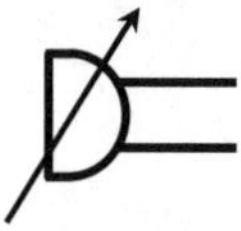

milliammeter

NAND gate

negative voltage
   connection

NOR gate

NOT gate

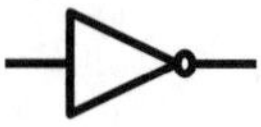

optoisolator

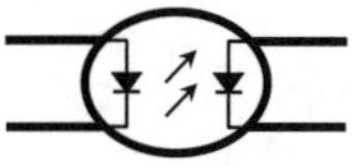

OR gate

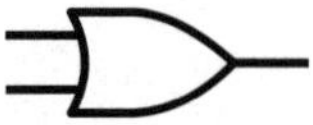

outlet, 2-wire, nonpolarized

outlet, 2-wire, polarized

outlet, 3-wire

outlet, 234-volt

plate, electron-tube

plug, 2-wire, nonpolarized

plug, 2-wire, polarized

plug, 3-wire

plug, 234-volt

plug, coaxial or phono

plug, phone, 2-conductor

plug, phone, 3-conductor

positive voltage connection

potentiometer

probe, radio-frequency

or

rectifier, gas-filled

rectifier, high-vacuum

rectifier, semiconductor

rectifier, silicon-controlled

relay, double-pole,
  double-throw

relay, double-pole,
  single-throw

relay, single-pole,
  double-throw

relay, single-pole,
  single-throw

resistor, fixed

resistor, preset

resistor, tapped

resonator

rheostat

saturable reactor

signal generator

solar battery

solar cell

source, constant-current

source, constant-voltage

speaker

switch, double-pole,
    double-throw

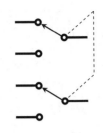

switch, double-pole,
    rotary

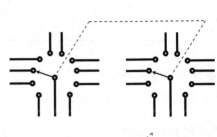

switch, double-pole,
    single-throw

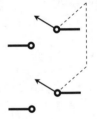

switch, momentary-contact

switch, silicon-controlled

switch, single-pole,
    double-throw

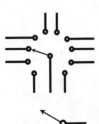

switch, single-pole, rotary

switch, single-pole, single-
    throw

terminals, general,
   balanced

terminals, general,
   unbalanced

test point
                TP

thermocouple
        or

transformer, air core

transformer, air core,
   step-down

transformer, air core,
   step-up

transformer, air core,
   tapped primary

transformer, air core,
   tapped secondary

transformer, iron core

transformer, iron core, step-down

transformer, iron core, step-up

transformer, iron core, tapped primary

transformer, iron core, tapped secondary

transformer, powdered-iron core

transformer, powdered-iron core, step-down

transformer, powdered-iron core, step-up

transformer, powdered-iron core, tapped primary

transformer, powdered-iron core, tapped secondary

transistor, bipolar, NPN

transistor, bipolar, PNP

transistor, field-effect,
    N-channel

transistor, field-effect,
    P-channel

transistor, MOS field-effect,
    N-channel

transistor, MOS field-effect,
    P-channel

transistor, photosensitive,
    NPN

transistor, photosensitive,
    PNP

transistor, photosensitive,
    field-effect, N-channel

transistor, photosensitive,
field-effect, P-channel

transistor, unijunction

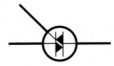

triac

tube, diode

tube, heptode

tube, hexode

tube, pentode

tube, photosensitive

tube, tetrode

tube, triode

unspecified unit or
   component

voltmeter

wattmeter

waveguide, circular

waveguide, flexible

waveguide, rectangular

waveguide, twisted

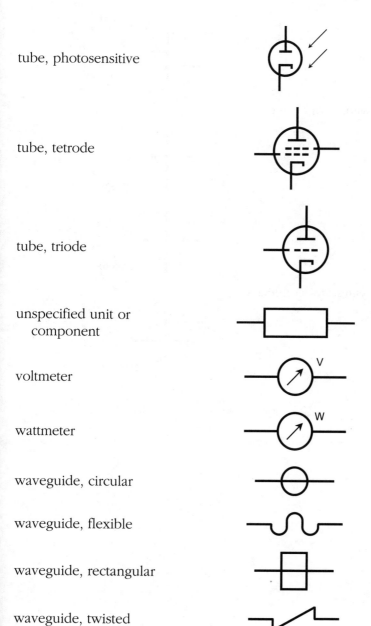

wires, crossing. connected

wires, crossing, not connected

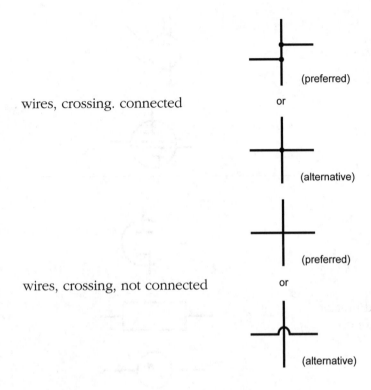

or

(preferred)

(alternative)

(preferred)

or

(alternative)

B

Resistor color codes

Some resistors have *color bands* that indicate their values and tolerances. You'll see three, four, or five bands around carbon-composition resistors and film resistors. Other resistors have enough physical bulk to allow for printed numbers that tell you the values and tolerances directly.

On resistors with *axial leads* (wires that come straight out of both ends), the first, second, third, fourth, and fifth bands are arranged as shown in Fig. B-1. On resistors with *radial leads* (wires that come off the ends at right angles to the axis of the component body), the colored regions are arranged as shown in Fig. B-2. The first two regions represent single digits 0 through 9, and the third region represents a multiplier of 10 to some power. (For the moment, don't worry about

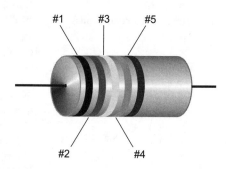

FIG. B-1. *Locations of color-code bands on a resistor with axial leads.*

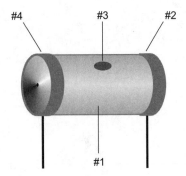

FIG. B-2. *Locations of color code designators on a resistor with radial leads.*

the fourth and fifth regions.) Table B-1 indicates the numerals corresponding to various colors.

Suppose that you find a resistor with three bands: yellow, violet, and red, in that order. You can read as follows, from left to right, referring to the table:

- Yellow = 4
- Violet = 7
- Red = ×100

You conclude that the rated resistance equals 4700 ohms, or 4.7 k.

Table B-1
The color code for the first three bands that appear on fixed resistors.
See text for discussion of the fourth and fifth bands

Color of band	Numeral (first and second bands)	Multiplier (third band)
Black	0	1
Brown	1	10
Red	2	100
Orange	3	1000 (1 k)
Yellow	4	10^4 (10 k)
Green	5	10^5 (100 k)
Blue	6	10^6 (1 M)
Violet	7	10^7 (10 M)
Gray	8	10^8 (100 M)
White	9	10^9 (1000 M or 1 G)

As another example, suppose you find a resistor with bands of blue, gray, and orange. You refer to Table B-1 and determine that:

- Blue = 6
- Gray = 8
- Orange = ×1000

This sequence tells you that the resistor is rated at 68,000 ohms, or 68 k.

If a resistor has a fourth colored band on its surface (#4 as shown in Fig. B-1 or B-2), then that mark tells you the tolerance. A silver band indicates ±10%. A gold band indicates ±5%. If no fourth band exists, then the tolerance is ±20%.

The fifth band, if any, indicates the maximum percentage by which you should expect the resistance to change after the first 1000 hours of use. A brown band indicates a maximum change of ±1% of the rated value. A red band indicates ±0.1%. An orange band indicates ±0.01%. A yellow band indicates ±0.001%. If the resistor lacks a fifth band, it tells you that the resistor might deviate by more than ±1% of the rated value after the first 1000 hours of use.

A competent engineer or technician always tests a resistor with an ohmmeter before installing it in a circuit. If the component turns out defective or mislabeled, you can prevent potential future headaches by following this simple precaution. It takes only a few seconds to check a resistor's ohmic value. If you skip that simple step, build a circuit, and then discover that it won't work because of some miscreant resistor, you might have to spend hours tracking it down!

Suggested additional reading

Frenzel, Louis E., Jr., *Electronics Explained*. Burlington, MA: Newnes/Elsevier, 2010.

Geier, Michael, *How to Diagnose and Fix Everything Electronic*. New York: McGraw-Hill, 2011.

Gerrish, Howard, *Electricity and Electronics*. Tinley Park, IL: Goodheart-Wilcox Co., 2008.

Gibilisco, Stan, *Electricity Demystified*, 2nd ed. New York: McGraw-Hill, 2012.

Gibilisco, Stan, *Electricity Experiments You Can Do at Home*. New York: McGraw-Hill, 2010.

Gibilisco, Stan, *Electronics Demystified*, 2nd ed. New York: McGraw-Hill, 2011.

Gibilisco, Stan, *Teach Yourself Electricity and Electronics*, 5th ed. New York: McGraw-Hill, 2011.

Gussow, Milton, *Schaum's Outline of Basic Electricity*, 2nd ed. New York: McGraw-Hill, 2009.

Horn, Delton, *Basic Electronics Theory with Experiments and Projects*, 4th ed. New York: McGraw-Hill, 1994.

Horn, Delton, *How to Test Almost Everything Electronic*, 3rd ed. New York: McGraw-Hill, 1993.

Kybett, Harry, *All New Electronics Self-Teaching Guide*, 3rd ed. Hoboken, NJ: Wiley Publishing, 2008.

Miller, Rex, and Miller, Mark, *Electronics the Easy Way*, 4th ed. Hauppauge, NY: Barron's Educational Series, 2002.

Mims, Forrest M., *Getting Started in Electronics*. Niles, IL: Master Publishing, 2003.

Morrison, Ralph, *Electricity: A Self-Teaching Guide*, 3rd ed. Hoboken, NJ: John Wiley & Sons, Inc., 2003.

Shamieh, Cathleen, and McComb, Gordon, *Electronics for Dummies*, 2nd ed. Hoboken, NJ: Wiley Publishing, 2009.

Slone, G. Randy, *TAB Electronics Guide to Understanding Electricity and Electronics*, 2nd ed. New York: McGraw-Hill, 2000.

Index

A

AC-to-DC converter, 13–14, 80–81
air-core inductor, 35–36
air-core transformer, 39
alligator clip, 118
AM radio receiver, 89–94
AM voice transmitter, 17–18
amateur radio station, 89
amplifier, 76–79, 85–94, 102
amplifier drive, 89
AND gate, 55
antenna, 63
antenna tuner, 89
arrows in flowcharts, 21
audio power amplifier, 85–89, 93, 102
avalanche effect, 134

B

battery, electrochemical, 54
bipolar transistor, 47–48
block diagram:
    definition of, 1–2
    examples and uses, 13–24
block- to schematic-diagram conversion, 14

breadboard, 113–117
buffer, 79
bulb, incandescent, 59, 61

C

capacitor:
    fixed, 31–32
    padder, 32
    polarized, 32
    trimmer, 32
    variable, 32–34
cat's whisker, 84
cell, electrochemical, 53–54, 59, 61
characters, 9
chassis ground, 35–36
chip, 17, 84
choke, 37
clip lead, 118
coaxial cable, 45–46
code practice oscillator, 96–103
coil, 63
component labeling, 66–71
component tolerance, 77
components, letter designations for, 68
components for experiments, 115

conductors:
    crossing, connected, 44
    crossing, not connected,
      43, 45
    single, 59, 61
conservation of current, 120
conservation of voltage, 123
conventional current, 58
"crystal radio" receiver, 84–85
current, conservation of, 120
current, conventional or
    theoretical, 58
current law, Kirchhoff's,
    119–122
cutoff frequencies, 103–104

D
decision block, 22–23
demodulator, 84
detector, 84
diode:
    semiconductor, 46–47, 63
    vacuum-tube, 49–50
directly heated cathode, 50
DPDT switch, 41
DPST switch, 41
drive, for amplifier, 88
dual triode vacuum tube, 52

E
earth ground, 45
electrochemical battery, 54
electrochemical cell, 53–54, 59,
    61
electron tube (*see* vacuum
    tube)
exclusive-OR gate, 55
experiments, components for,
    115
external capacitance effects, 32

F
farad, 30–31
ferromagnetic material, 37
field-effect transistor, 48–49
field-strength meter, 62–64
filament in vacuum tube,
    49–50
first law, Kirchhoff's, 120
fixed capacitor, 31–32
fixed inductor, 35–38
flashlight:
    single-cell, 58–60
    two-cell, 60–62
flowchart, 18–24
flux density, magnetic, 37
forward bias, 133–134
forward breakover voltage,
    134
full-wave bridge rectification,
    69
functional diagram, 13–16

G
galena, 84
ganged capacitors, 34
Gunn diode, 46

H
hand key, 42–43
hard wiring, 7
hardware, 23
henry, 35
heptode vacuum tube, 52–53
hexode vacuum tube, 52–53
highpass filter, 103–104

I
incandescent bulb, 59, 61
indirectly heated cathode,
    49–50

inductance-capacitance circuit,
   92, 103–104
inductor:
      air-core, 35–36
      fixed, 35–38
      laminated-iron-core, 37
      powdered-iron-core, 38
      solid-iron-core, 37
      tapped, 35–36, 38
      variable, 36, 38
input and output programming
   symbol, 20
integrated circuit, 17, 84
interconnections, schematic,
   6–8
intermediate junction program
   symbol, 20
International Morse code, 96
iron-core inductor, 37–37
iron-core transformer, 39

J–K
jumper wire, 118

key for sending Morse code,
   42–43
Kirchhoff, Gustav Robert, 120
Kirchhoff's current law, 119–122
Kirchhoff's first law, 120
Kirchhoff's second law, 123
Kirchhoff's voltage law, 123–125

L
L network, 94
labeling of components, 66–71
laminated-iron-core inductor,
   37
laminated-iron-core transformer,
   39
lamps, mismatched, 137–144

language:
      sign, 10
      visual, 8–11
LC circuit, 92, 103–104
letter designations for
   components, 68
load resistor, 130
logic gates, 54–55
logical inverter, 54–55
lowpass filter, 103–104

M
magnetic flux density, 37
microammeter, 63
microfarad, 31
microhenry, 35
millihenry, 35
mismatched lamps, 137–144
Morse code key, 42–43
multicontact switch, 40–42
multimeter, 74

N
N-type semiconductor, 132
NAND gate, 55
nanohenry, 35
nonferromagnetic material, 37
nonpolarized device, 31
NOR gate, 55
NOT gate, 54–55

O
off-page connection flowchart
   symbol, 20
ohm, unit of resistance, 25
OR gate, 55
oscillator, 79, 96–103

P
P-N junction, 134

P-type semiconductor, 132
padder capacitor, 32
page breaks, 91–94
parallel resistors, 70–71
peak inverse volts, 132
pentagrid converter, 53
pentode vacuum tube, 51–52
phase coincidence, 97
phase inversion, 97
pi network, 94–95
pi-L network, 95
picofarad, 31
pictorial diagram, definition of, 1
pictorial versus schematic
   diagrams, 11
polarized capacitor, 32
polarized device, 31
pole, switch, 40
positive-ground system, 97
potentiometer, 26–30
powdered-iron-core inductor,
   38
powdered-iron-core
   transformer, 39
power amplifier, audio, 85–89
power supply, 16–17, 64–65,
   80–81, 97–100
preamplifier, 85–87, 91–94
prewritten program symbol,
   20
process paths, 22–24
processing operation programming symbol, 20
program decision symbol, 20
program flowchart, 19–24
program modification symbol,
   20
punched-card program
   flowchart, 21–22

push-pull circuit, 87–89

R
radio receiver, 89–94
rectifier diode, 46–47
resistor:
   carbon composition, 26
   film type, 27
   fixed-value, 26–27
   load, 130
   variable, 28–30
   wirewound, 27, 29
resistors in parallel, 70–71
resonant frequency, 85
reverse bias, 133–134
rheostat, 26–29
ripple, 65
roller inductor, 95
rotary switch, 42
rotor of variable capacitor, 33

S
schematic-and-block combinations, 78–81
schematic diagram, definition
   of, 1–4
schematic interconnections,
   6–8
schematic symbology, purpose
   of, 4–6
schematic- to block-diagram
   conversion, 14–15
schematic versus pictorial
   diagrams, 11
second law, Kirchhoff's, 123
sign language, 10
silicon-controlled rectifier, 47
single-cell flashlight, 58–60
software, 24

solid-iron-core inductor, 37
SPDT switch, 40
speech processor, 89–90
SPST switch, 40
start programming symbol, 20
stator of variable capacitor, 33
stop programming symbol, 20
straight key, 42–43, 96
strobe light circuit, 15–16,
    108–111
switch:
    DPDT, 40–41
    DPST, 41
    multicontact, 40–42
    rotary, 42
    SPDT, 40
    SPST, 40, 61

T

tapped inductor, 35–36, 38
telegraph key, 96
test points, 75–76
tetrode vacuum tube, 51
theoretical current, 58
throw, switch, 40
tolerance, component, 77
transformers, all types, 39
transistor:
    bipolar, 47–48
    field-effect 48
trimmer capacitor, 32
triode vacuum tube, 50–51
troubleshooting, 71–75, 103

tube (*see* vacuum tube)
tuner, in radio receiver, 92–94
twin-T oscillator, 96–103
two-cell flashlight, 60–62

U–Z

vacuum tube:
    diode, 49–50
    dual triode, 52
    heptode, 52–53
    hexode, 52–53
    pentode, 51–52
    tetrode, 51
    triode, 50–51
varactor diode, 46
variable capacitor, 32–34
variable inductor, 36, 38
visual language, 8–11
voice transmitter, 17–18
volt-ohm-milliammeter, 74
voltage, conservation of, 123
voltage divider, 125–132
voltage doubler, 69–70
voltage law, Kirchhoff's,
    123–125
voltage reducer, 132–137
voltage regulation, 69

wire wrapping, 117–118

XOR gate, 55

Zener diode, 46, 134